Efficiency

Improving the Performance of
Your SAS® Applications

Robert Virgile

Comments or Questions?

The author assumes complete responsibility for the technical accuracy of the content of this book. If you have any questions about the material in this book, please write to the author at this address:

SAS Institute Inc.
Books by Users
Attn: Robert Virgile
SAS Campus Drive
Cary, NC 27513

If you prefer, you can send e-mail to sasbbu@sas.com with "comments for Robert Virgile" as the subject line or you can fax the Books by Users program at (919) 677-4444.

The correct bibliographic citation for this manual is as follows: Virgile, Robert, *Efficiency: Improving the Performance of Your SAS® Applications*, Cary, NC: SAS Institute Inc., 1998. 256 pp.

Efficiency: Improving the Performance of Your SAS® Applications

Copyright © 1998 by SAS Institute Inc., Cary, NC, USA.

ISBN 1-58025-228-1

SAS Institute Inc., SAS Campus Drive, Cary, North Carolina 27513.

1st printing, June 1998

Note that text corrections may have been made at each printing.

The SAS® System is an integrated system of software providing complete control over data access, management, analysis, and presentation. Base SAS software is the foundation of the SAS System. Products within the SAS System include SAS/ACCESS®, SAS/AF®, SAS/ASSIST®, SAS/CALC®, SAS/CONNECT®, SAS/CPE®, SAS/DB2™, SAS/DMI®, SAS/EIS®, SAS/ENGLISH®, SAS/ETS®, SAS/FSP®, SAS/GEO™, SAS/GIS®, SAS/GRAPH®, SAS/IML®, SAS/IMS-DL/I®, SAS/INSIGHT®, SAS/IntrNet™, SAS/LAB®, SAS/MDDB®, SAS/NVISION®, SAS/OR®, SAS/PH-Clinical®, SAS/QC®, SAS/REPLAY-CICS®, SAS/SECURE®, SAS/SESSION®, SAS/SHARE®, SAS/SHARE*NET™, SAS/SPECTRAVIEW®, SAS/SQL-DS™, SAS/STAT®, SAS/TOOLKIT®, SAS/TUTOR®, and SAS/Warehouse Administrator™ software. Other SAS Institute products are SYSTEM 2000® Data Management Software, with basic SYSTEM 2000, CREATE™, Multi-User™, QueX™, Screen Writer™, and CICS interface software; InfoTap®, JMP®, JMP IN® and JMP Serve®, and StatView® software; SAS/RTERM®, the SAS/C® Compiler, Budget Vision™, Campaign Vision™, CFO Vision™, Emulus®, Enterprise Miner™, Enterprise Reporter™, HR Vision™ software, IT Charge Manager™ software, and IT Service Vision™ software, MultiVendor Architecture™ and MVA™, MultiEngine Architecture™ and MEA™, Risk Dimension™, SAS InSchool™, Scalable Performance Data Server™, Video Reality™, Warehouse Viewer™ are trademarks or registered trademarks of SAS Institute Inc. SAS Institute also offers SAS Consulting® and SAS Video Productions® services. Authorline®, Books by Users℠, The Encore Series®, ExecSolutions®, JMPer Cable®, Observations®, SAS Communications®, sas.com™, SAS OnlineDoc™, SAS Professional Services™, the SASware Ballot®, SelecText™, and Solutions@Work™ documentation are published by SAS Institute Inc. The SAS Video Productions logo, the Books by Users SAS Institute's Author Service logo, the SAS Online Samples logo, and The Encore Series logo are registered service marks or registered trademarks of SAS Institute Inc. The Helplus logo, the SelecText logo, the Video Reality logo, the Quality Partner logo, the SAS Business Solutions logo, the SAS Rapid Warehousing Program logo, the SAS Publications logo, the Instructor-based Training logo, the Online Training logo, the Trainer's Kit logo, and the Video-based Training logo are service marks or trademarks of SAS Institute Inc. All trademarks above are registered trademarks or trademarks of SAS Institute Inc. in the USA and other countries. ® indicates USA registration.

The Institute is a private company devoted to the support and further development of its software and related services.

Other brand and product names are registered trademarks or trademarks of their respective companies.

Table of Contents

iv

Acknowledgment

Many individuals must work extremely hard to turn a manuscript into a book. From technical reviewers to editors to production staff, those whose names do not appear still made a large contribution. Special thanks go to Judy Whatley, who coordinated my requests (both reasonable and otherwise) among many departments; to Margaret Crevar who managed to run my test programs on four operating systems; and to my wife Paula, who had to endure long stretches of my burning the candle at both ends.

Preface

Key Features of This Book

How does this book differ from other material on computer efficiency? Four features stand out:

- It concentrates on tasks that programs perform regularly, such as reading data, sorting data, and summarizing data.

- It rates the savings, so you can select the more important material to learn first.

- It demonstrates how to write test programs that you can run in your own environment.

- It covers programming technique and strategies, not just syntax.

Focusing on Common Programming Tasks

This book concentrates on tasks that programs perform regularly, such as reading data, sorting data, and summarizing data. I have devoted one chapter apiece to these topics and to other common tasks as well. On the other hand, I have avoided topics that occur infrequently. For example, the NOTSORTED option in PROC FORMAT stores format ranges in the same order as they appear in the VALUE statement:

```
PROC FORMAT;
VALUE F1_  (NOTSORTED)
      1='A'  2='B'  3='C'  4='D'  5='E'  6='F';
VALUE F2_
      1='A'  2='B'  3='C'  4='D'  5='E'  6='F';
```

If you attempt to look up a matching value with the PUT function, the software performs a sequential search when through the F1_ format (faster if the first argument to the PUT function is 1) and a binary search through the F2_ format (faster for most values):

```
NEWVAR1 = PUT(1, F1_.);
NEWVAR2 = PUT(1, F2_.);
```

So when you know that 1 is usually the incoming value, you can speed up the search process using the NOTSORTED option. Besides being rare, however, this option is unnecessary. If you know the frequency of the incoming data values, you can code this way:

```
IF X=1 THEN NEWVAR='A';
ELSE NEWVAR = PUT(X, F2_.);
```

If local programming standards or complex macro code force you to use a single statement to perform the recoding, you can still get the best of both worlds by creating two formats and having one format reference the other:

```
PROC FORMAT;
VALUE F1_ 1='A'
          Other=[F2_];
VALUE F2_ 2='B' 3='C' 4='D' 5='E' 6='F';
```

Now a single statement does the job:

```
NEWVAR = PUT(X, F1_.);
```

It performs a sequential search through F1_ followed by (if needed) a binary search through F2_. Enough of the esoteric! While some topics in the book, such as random sampling, may fall outside the scope of your work, one glance at the table of contents will tell you that the book focuses on common programming tasks.

Rating the Savings

The book rates the savings according to the quantity of CPU time or storage space saved, so you can select the more important material to learn first. Each tool or technique receives a rating of one to three stars. Take these ratings with a grain of salt, however, because small savings can add up. If you notice a one-star topic that applies quite frequently in your programs, upgrade it a notch to two stars.

Test Programs

A series of test programs measures the savings in CPU time. Chapter 10 reports the results on four common operating systems. If you would like, you can download these programs and test them on your own hardware and release of the software. See the inside back cover of this book for details.

By rerunning these programs, you can verify that your savings match the savings reported in the book. In addition, as new releases of the software become available, these programs can test whether the old techniques still work. In many situations, you may wish to extend the tests in this book, writing your own test programs. In some cases, savings may fluctuate depending on characteristics of your data. In other cases, the number of testable variations is too large to incorporate in a book. From time to time, you may dream up your own ideas about potential efficiencies. Consider this program as an imaginative possibility:

```
PROC FORMAT;
VALUE    $VAL
'a'='A'  'b'='B'  'c'='C'  'd'='D'  'e'='E'  'f'='F';
INVALUE $INVAL
'a'='A'  'b'='B'  'c'='C'  'd'='D'  'e'='E'  'f'='F';
```

The program creates both an informat and a format that produce the same character-to-character translations. If you think about it, either the format or the informat could be used to recode the lowercase values to the uppercase values:

```
NEWVAR1 = PUT  ('a', $VAL.);
NEWVAR2 = INPUT('a', $INVAL.);
```

Both NEWVAR1 and NEWVAR2 receive a value of A. Just for the record, the PUT function works faster. This is another somewhat esoteric topic, not included later in the book. But the real lesson is that you will find many topics through your own programs and your own ingenuity where you may want to test one programming technique against another. The final chapter in this book will guide you in creating test data and extending test programs beyond the scope of this book.

Programming Techniques and Strategies

The book addresses programming techniques and strategies, not just syntax. For example, Chapter 6 "Summarizing Data," also explores resummarizing a summary data set. Chapter 2, "Reading Data," talks about sampling techniques. Chapter 2 also explains WHERE vs. IF for subsetting, not only considering which is faster, but also explaining other programming considerations that come into play that might make the slower tool the better tool.

In short, this book focuses on the most common types of programming applications. It shows you how to reduce the resources for running programs by modifying both SAS statements and programming technique and strategies.

CHAPTER 1

Introduction

What Is Efficiency?

To be efficient, a program must:

- use the least costly set of resources, and

- accomplish the programming objectives.

Both criteria sound straightforward. Neither one is quite so simple.

Evaluating the Cost of Resources

Consider the first issue, using the least costly set of resources. Each computer system will include a variety of hardware and services. For example, you may pay dearly for:

- disk space

- CPU time

- I/O *- input- output / key / More groups*

- technical support

- tape drives

- turnaround time.

Even if your fee schedule does not list all of these items, you may pay for them indirectly. For example, when using a PC, there is no direct charge for CPU time. However, the longer a program runs, the longer you must sit and wait for it to finish. On a mainframe, you may not pay for using tape drives. However, jobs that use many tape drives may have to wait until the tape drives become available, perhaps running overnight.

An abundant resource at one site might be scarce at another site. At one site, disk space might be plentiful and CPU time costly. At another site, the reverse might be true. One site might provide a Help Desk with a senior programmer who knows the SAS System extremely well. At another site, you might be lucky to get a manual. Under different cost structures, you might use entirely different programming techniques because the relative costs of the resources change.

Consider three ways of storing this entirely fictitious data. Structure #1 is the most flexible for analysis purposes:

Variables:	TEAM	WINS	YEAR	
Values:	RED SOX	80	1997	
	RED SOX	80	1998	
	RED SOX	85	1999	
	TIGERS	105	1997	
	TIGERS	90	1998	
	TIGERS	80	1999	
	YANKEES	80	1997	
	YANKEES	90	1998	
	YANKEES	110	1999	

by year

Structure #2 might work best for printing the data or converting it to spreadsheet form:

Variables:	TEAM	WIN97	WIN98	WIN99
Values:	RED SOX	80	80	85
	TIGERS	105	90	80
	YANKEES	80	90	110

Structure #3 is best for generating plots on an x-y axis:

Variables:	RED_SOX	TIGERS	YANKEES	YEAR
Values:	80	105	80	1997
	80	90	90	1998
	85	80	110	1999

Which structure is best? As you sit down to discuss the matter with your boss, you might say, "It really doesn't matter which form we use. I can write a program to convert from one form to another whenever it becomes necessary." And your boss might reply, "I know you can. But I have 10 other users who can't. They have to do the research, and they need the data available in all three forms. So store it all three ways or else they'll be bothering you for help all the time." And you say, "But, but, but..." And the boss says, "No, no, no." Who's right?

Grudgingly, you may have to admit that the boss might be right. The bottom line is that there is no one right answer. It might be cheapest to store the data three times instead of once, and the best strategy might differ from one project to the next. Besides the dollar cost for storage space, take into consideration

- the skill level of the users

- the amount of free time you have to help out the users

- the number of expected changes to be made to the data in the future

- whether you, as the senior programmer, have the skill to store the data in one form, and store the other forms as views rather than data sets.

Does the Program Work?

What about the second criterion for efficiency: Does the program accomplish its objectives? This too sounds like a clear-cut issue; the program works or it doesn't. Once again, gray areas abound. Consider the possibilities below.

When the program contains a syntax error, it certainly did not accomplish its objectives. However, be sure to check the SAS log. The existence of output does not guarantee an error-free program. Perhaps the first half of the program produced valid output, while a later SAS step contained an error. Also, if your method of running the SAS System produces .LOG and .LST files as output, the existence of a .LST file does not guarantee zero syntax errors. It is quite possible that the latest run of the program contained many errors, and that the .LST file was produced by running an earlier version of the same program.

When the program contains a logic error, the output can be wrong. Sometimes, notes in the SAS log will provide a clue. Messages about numeric to character conversion, lost card, or variables being uninitialized sometimes indicate a logic error. Still, many logic errors, such as faulty IF / THEN / ELSE logic, produce no messages to help you.

A program can be too complex. Have you ever seen a 200 line program with no comment statements, no spacing or indention, and complex programming techniques? It may work perfectly. But nobody could possibly know this because nobody can figure out how the program produces its results.

Perhaps the program works, but it has to run overnight because it ties up key resources such as an important data set or too many tape drives. The longer you have to wait for the results, the less efficient the program.

Perhaps the program would work, but it is too expensive to run.

Finally, what if the program works, but it takes a lot of effort to interpret the results? A more efficient program would clearly display what the user needs to know.

Weighing the Benefits and Costs

Many considerations go into labeling a program as efficient or inefficient, but they all fall into these two basic categories: the efficient program uses the least costly set of resources, and the program accomplishes the programming objectives. When this book presents a technique that might speed up your program, you still must decide whether it makes sense for you to apply the technique. When evaluating a technique, consider these issues:

- How much does this programming technique save? If your program processes a small data set, you shouldn't spend three hours of your time to cut a program's CPU time in half. Half of nothing is nothing.

- How often does the program run? If you save 5% of the CPU time for a program which runs daily, that can add up over the course of a year.

- Do you feel comfortable with the programming technique? Practice makes perfect. But if you are uncertain whether the results are accurate, the program is useless.

- Do you have the time to spend investigating alternative programming techniques? Perhaps your workload is too heavy to worry about trying something new. Perhaps the program results are needed quickly.

My bottom line recommendation is this: learn some new techniques for improved efficiency and make them habits. Over time, incorporate more of these into your programs.

Rating the Savings

To help you evaluate the importance of various sections, this book rates the topics and programming techniques, using one to three stars:

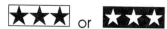 or

Saves the most CPU time or storage space

 or

Medium savings

 or

Smallest savings

Black stars vs. white stars reveal my assessment of whether the technique applies frequently (black stars) or infrequently (white stars). Naturally, you may find that frequency in your programs differs from my assessment.

Two additional icons cover special situations:

The savings fluctuate widely, depending upon your operating system or the characteristics of your data. The savings can range from large to nonexistent to negative.

This technique does not save. It may have, under previous releases of the software.

If you notice that a technique would be useful in many of your programs, add another star. These ratings are general guidelines only, to help you prioritize which material is most valuable to learn first.

Be careful when interpreting the percentage savings offered by alternative statements (discussed in Chapter 10). The percentage savings in CPU time represents the savings for that one statement, not the savings for the entire DATA step. Also, note that saving 10% of the CPU time when the bill is $200 is more valuable than saving 25% when the bill is $10.

The Focus of This Book

This book focuses on getting programs to run faster. Most of the book addresses reducing CPU time (Chapters 2 through 7), with one chapter on saving storage space (Chapter 8). Often, I made an executive decision to place a topic in one chapter or another. For example, the CLASS statement in PROC MEANS works on unsorted data. By replacing a BY statement, you eliminate the need for PROC SORT. I placed this topic in Chapter 5 (sorting) not in Chapter 6 (summarizing data).

For the most part, the book does not distinguish between saving CPU time vs. saving on I/O. Chapter 9 addresses I/O briefly, as well as other general efficiency items that do not fall neatly into earlier chapters. Still, not even Chapter 9 delves into system options that affect memory management or special situations such as extremely large data sets. The intent is to cover programming tools and techniques that come into play for the vast majority of applications programmers.

2

Finally, Chapter 10 displays the savings generated by the test programs used in writing this book. You can download copies of these programs to verify that the savings on your hardware and the release of the software match the results in this book. This chapter also explains how to generate test data and write programs to test your ideas on efficiency.

CHAPTER 2

Reading Data

Overview

Most programs read data. And the CPU time needed to read data (particularly raw data) can far outweigh that for manipulating the data. As a result, the techniques in this chapter can generate substantial savings.

This chapter contains three sections. Section 1 examines DATA step techniques for reading raw data. Section 2 examines techniques for reading SAS data, whether in a DATA or PROC step. Section 3 covers more general concepts, such as sampling methods and beating the DATA step loop.

Legend for Icons

CPU time or storage space
Black stars = technique applies frequently
White stars = technique applies infrequently

3 black or white stars - maximum savings
2 black or white stars - medium savings
1 black or white star - smallest savings

Fluctuating savings
Savings can range from large to nonexistent to negative, depending on your operating system or data characteristics.

Archaic
No savings under current release.

Section 1: Reading Raw Data

When reading from raw data, the most important CPU time savers are:

- Read only the needed variables. ✓

- When subsetting, read key variables first. ✓

- Choose faster forms of the INPUT statement.

Secondary techniques include:

- Replace data manipulation with informats. ✓

- Read multiple files in a single DATA step. ✓

Let's begin with the biggest savers of CPU time.

Read Only the Needed Variables

Reading from raw data is expensive. Analyze which variables are needed and read those only:

```
INPUT   NAME    $    1- 20
        RANK    $   71- 80
        SERIAL  $ 251-260;
```

If your program does not need the data from columns 21 through 70, don't read in those variables. This is an easy step to overlook if you copy an existing program that includes a longer INPUT statement.

NAME $20.   INFORMAT
XXX $49.   die bequeable
RANK $10.
YYY $170.   MUSAIYHG
SERIAL $10.

Read Key Variables First When Subsetting from Raw Data

The INPUT statement does more than map the layout of the raw data. It contains a series of instructions that reads in data. This fact impacts the DATA step below, where the userid XYZ1234 appears in fewer than 1% of the raw data lines. For the remaining 99% of the observations, the DATA step reads in the other five variables unnecessarily:

```
DATA _1_USER;
INFILE HUGETAPE;
INPUT USERID  $ 1-7
      DATASET $ 8-51
      LRECL     52-56
      RECFM     57-59
      BLKSIZE   60-64
      PACK    $65-70;
IF USERID='XYZ1234';
```

Instead of reading all variables for all observations, the DATA step should:

- Read just the variable(s) needed to subset the observations.

- Apply the subsetting IF statement.

- If needed, read in the remaining variables.

The revised DATA step:

```
DATA _1_USER;
INFILE HUGETAPE;
INPUT USERID  $ 1-7 @;
IF USERID='XYZ1234';
INPUT DATASET $ 8-51
      LRECL     52-56
      RECFM     57-59
      BLKSIZE   60-64
      PACK    $65-70;
```

Normally, each INPUT statement reads its variables and then releases the line of raw data. But the trailing @ holds the current line of raw data, enabling a later INPUT statement to read additional variables from the same data line. (The DATA step may never execute a later INPUT statement, as would happen when USERID is not XYZ1234. In that case, the software eventually releases the line of data when the sub-setting IF deletes the observation, causing the program to return to the DATA statement.)

The savings can be enormous. In this case, reading one variable instead of six for 99% of the data lines would save over 80% of the CPU time.

Choose Faster Forms of INPUT

The INPUT statement supports different styles: list input, column input, formatted input, and named input. Depending on the structure of the raw data, you may have to use one style or another. But in most cases, you have choices. Remember, the same INPUT statement can mix and match, reading some variables in one style and other variables in another. Let's examine which tools work faster.

Choose Faster Forms of INPUT: Avoid List Input

These two INPUT statements generate identical results:

```
INPUT @11 AGE 2. ;
INPUT AGE 11-12;
```

As long as the data meet two conditions (column 12 or 13 must always be blank, and columns 11 and 12 cannot both be blank), the next INPUT statement also generates the same results:

```
INPUT @11 AGE;
```

However, the third one runs slower. The first two examine exactly two characters from the raw data line, while the last must examine at least two characters and possibly three. When columns 11 and 12 both contain digits, the last INPUT statement also examines column 13 to verify that it is blank. Therefore, when adding a two-digit variable to an existing INPUT statement, expect the following result. Adding the variable using list input will add approximately 50% more CPU time then the other forms of input.

Choose Faster Forms of INPUT: $CHAR Informats

These INPUT statements produce similar results:

```
INPUT @1 ZIPCODE1 5.;
INPUT @1 ZIPCODE2 $5.;
INPUT @1 ZIPCODE3 $CHAR5.;
```

ZIPCODE1 is numeric, while ZIPCODE2 and ZIPCODE3 are character. For reporting purposes, the character variables are slightly easier to work with because they preserve any leading zeroes. For other purposes, the variables are interchangeable. Although ZIP codes contain digits, they are categorical by nature. You will never need to add ZIP codes together or multiply a ZiP code by another variable. All three variables can be stored in roughly the same number of bytes. However, in terms of speed, the three INPUT statements are quite different.

The first INPUT statement creates ZIPCODE1 as a numeric variable. Because the SAS System stores all numerics in binary floating point, the software must take extra steps to convert the digits in the raw data to the proper numeric form. The other two INPUT statements skip those extra steps, copying the characters from columns 1 through 5. However, the $5. informat automatically left-hand justifies the contents of columns 1 through 5. The $CHAR5. informat copies the raw data as is, without left-hand justifying. Even when your raw data are originally left-hand justified, the $CHAR5. informat still runs faster than the $5. informat. The $5. informat must examine column 1 to see whether it is blank or not.

Test results on four operating systems reveal that the $CHAR5. informat always runs faster than the $5. informat. Surprisingly, however, the 5. informat varies from slowest to fastest, depending on the operating system. See Chapter 10 for details.

Choose Faster Forms of INPUT: Reading Consecutive Fields

As the SAS System reads through a line of raw data, it tracks its current position within that line. These two INPUT statements, therefore, produce identical results:

```
INPUT @1 NAME $20. @21 AGE 2.;
INPUT NAME $20. AGE 2.;
```

When beginning to read the data line, the software begins automatically at column 1. There is no need to add @1 to position the software to that spot. Having just read NAME from columns 1 through 20, the software is automatically positioned at column 21. Again, there is no need to add @21 to position the software to that spot. By carrying out those instructions (@1 and @21), the program performs extra work.

When reading consecutive variables, comment out the starting columns:

```
INPUT /* @ 1 */ NAME $20.
      /* @21 */ AGE    2.;
```

Retaining the commented out starting columns can help in two ways. First, they serve as documentation, letting you verify that your informats are correct. By adding the informat length to each variable's starting position, you should find the starting position for the next variable. Second, the starting columns allow you to copy the program as a starting point for later programs. If a later program requires only a handful of variables, uncomment the starting positions, letting you remove variables while still creating an accurate INPUT statement.

Reading Raw Data: Secondary Considerations

A few additional techniques can save smaller amounts of CPU time when reading raw data.

Replace Data Manipulation with Informats

Informats transform values as the INPUT statement reads the data. You can eliminate later data manipulation with the right informat. A simple case:

```
INPUT @1 CHARDATE $CHAR6.;
SASDATE = INPUT(CHARDATE,YYMMDD6.);
```

value of   INFORMAT

A faster program would eliminate the INPUT function, using the YYMMDD6. informat to perform the data manipulation:

```
INPUT @1 SASDATE YYMMDD6.;
```

yymmddy

In sample test runs, the single statement ran faster than the original combination of two statements, on all four operating systems. The savings in CPU time ranged from 6% to 26%.

User-defined informats can replace a broader range of data manipulation statements. Assuming that the raw data always contain the word MALE or FEMALE, this program recodes the raw data values:

```
INPUT @1 GENDER $CHAR6.;
IF GENDER='MALE' THEN TYPE=1;
ELSE TYPE=2;
```

type   *FEMALE - 6.*

MALE } VALUE
FEMALE } of VARIABLE

Instead, you can also define your own informat capable of recoding the incoming data values:

```
PROC FORMAT;
INVALUE $GENDER 'MALE'=1 'FEMALE'=2;
```

Now the subsequent INPUT statement can apply the informat:

```
INPUT @1 TYPE $GENDER6.;
```

TYPE becomes a numeric variable, with values of 1 and 2, just as it was in the original DATA step. However, this INPUT statement takes much longer than the original! In the original version, all components – the IF/THEN statements and the $CHAR6. informat – run very quickly. In the new program, the user-defined informat takes at least 50% more CPU time.

In this final example, consider raw data containing dates written by a traditional programming language such as COBOL. These dates appear in YYMMDD form, with missing values represented by 000000. Without an informat, a DATA step might read the data using:

```
INPUT @1 CHARDATE $CHAR6.;
IF STRING NE '000000' THEN DATE = INPUT(STRING, YYMMDD6.);
```

However, try creating this informat first:

```
PROC FORMAT;
INVALUE MISSD 0=. OTHER=[YYMMDD6.];
```

Now the program can read the date values directly:

```
INPUT @1 DATE MISSD6.;
```

No tests were run on this variation.

Read Multiple Raw Data Files in One Step

When your program must read several raw data files and concate-
nate the results, your first thought might be to read each file
separately:

```
DATA SOURCE1;
INFILE RAW1;
INPUT NAME $ 1-20   RANK $ 21-23   SERIAL 32-40;

DATA SOURCE2;
INFILE RAW2;
INPUT NAME $ 1-20   RANK $ 21-23   SERIAL 32-40;

DATA SOURCE3;
INFILE RAW3;
INPUT NAME $ 1-20   RANK $ 21-23   SERIAL 32-40;

DATA ALL3;
SET SOURCE1 SOURCE2 SOURCE3;
```

However, this approach performs extra work. The first three DATA steps
read and output every observation. The final DATA step reads and
outputs every observation over again. The final DATA step does not
double the workload because reading SAS data sets is faster than
reading raw data. However, the final step is unnecessary because
one DATA step can read and output all the observations from all three
sources of data. A faster program would read each source in a
DO loop:

```
DATA ALL3;
INFILE RAW1 END=EOF1;
DO UNTIL (EOF1);
    INPUT NAME $ 1-20 RANK $ 21-23 SERIAL 32-40;
    OUTPUT;
END;
INFILE RAW2 END=EOF2;
DO UNTIL (EOF2);
    INPUT NAME $ 1-20 RANK $ 21-23 SERIAL 32-40;
    OUTPUT;
END;
INFILE RAW3 END=EOF3;
DO UNTIL (EOF3);
    INPUT NAME $ 1-20 RANK $ 21-23 SERIAL 32-40;
    OUTPUT;
END;
```

This program contains a hidden drawback. If you intend to perform data manipulation, you must add an ELSE statement for each IF/THEN statement. For example, if this statement were to appear inside a DO loop, it would produce the wrong result:

```
IF NAME='Ivan' THEN NICKNAME='The Terrible';
```

Once the program encounters a single Ivan, all values of NICKNAME from that point forward will be The Terrible. The DATA step automatically resets NICKNAME to missing as the software passes through the DATA statement. However, when the DATA step reads all observations in a DO loop, it never passes through the DATA statement between observations. Instead, the program must reset NICKNAME to missing:

```
ELSE NICKNAME=' ';
```

Section 2: Reading from SAS Data Sets

When reading from SAS data sets, efficiency techniques include:

- Read just the needed variables. Although this applies to reading raw data as well, the techniques are quite different. As a result, this item appears in both Section 1 and Section 2.

- Choose the proper tool for subsetting observations.

- When processing a "wide" data set several times, subset the variables first.

Read Only the Needed Variables

When reading from SAS data sets, the KEEP= and DROP= data set options control the variables you process:

```
DATA MILK (KEEP=NAME CUPS);
SET HISTORIC.COWS (KEEP=NAME PINTS QUARTS);
CUPS = 2*PINTS + 4*QUARTS;
```

In the SET statement, the KEEP= data set option limits the variables read from HISTORIC.COWS. This operates differently than KEEP= in the DATA statement, which limits the variables output to MILK. Both steps have an impact, but in different ways.

Clearly, MILK ends up with two variables whether KEEP= appears in the SET statement. However, HISTORIC.COWS might contain 500 variables. Without KEEP=, the SET statement would read in all 500. That would be the major cost of the DATA step. Calculating CUPS and outputting two variables would be very fast by comparison. So even though KEEP= appears in the DATA statement, it saves in the SET statement as well.

Cases That Don't Benefit

Even PROC steps can add the KEEP= data set modifier. However, doing this does not help. Intuitively, KEEP= should make no difference in procedures. In PROC PRINT and PROC MEANS, for example, the VAR statement (and others, such as BY and ID) list all needed variables. Therefore, KEEP= should make no difference in these two PROC PRINTs:

```
PROC PRINT DATA=TEMP;
VAR VAR1 VAR5 VAR8 VAR98;

PROC PRINT DATA=TEMP (KEEP=VAR1 VAR5 VAR8 VAR98);
VAR VAR1 VAR5 VAR8 VAR98;
```

In this case, intuition proves to be correct. When other statements determine which variables get processed, a procedure cannot gain by adding the KEEP= or DROP= data set modifiers.

PROC SORT never adds statements to determine which variables to process. Therefore, KEEP= does affect PROC SORT. Chapter 5 presents those details.

Observations: Read What You Need

To determine the fastest method for subsetting observations, you need to know some of the theory behind the alternative statements. Although the DELETE statement can be used in any DATA step, the WHERE statement can only be used with SAS data sets. Therefore, this topic belongs in the current section concerning SAS data sets. Section 3 in this chapter addresses testing programs on a subset or sample of the data.

Subsetting with IF vs. DELETE vs. WHERE

The DATA step supports the subsetting IF, DELETE, and WHERE statements for subsetting. Which will be faster? <u>The subsetting IF and DELETE statements are equally efficient.</u> Simply pick the statement that is easier to program or easier to interpret. This example assumes that AGE takes on integer values:

```
DATA SENIORS;
   SET EVERYONE;
   IF AGE > 80;
vs.
   IF AGE < 81 THEN DELETE;
```

When comparing IF vs. WHERE, however, even the simplest of DATA steps find one statement running faster than the other. Consider how IF and WHERE perform the subsetting in this DATA step:

```
DATA SENIORS;
   SET EVERYONE;
   WHERE /* vs. IF */ AGE > 80;
```

When subsetting with IF, the DATA step reads in every observation from EVERYONE. Those that fail the subsetting IF test get deleted. Subsetting with WHERE works differently, however. The WHERE statement examines each observation before reading it into the Program Data Vector. It reads only the needed observations. In theory, this should make the WHERE statement faster. After all, the program doesn't read in an observation only to delete it with the next statement. In practice, the process of looking through the observations to determine which are needed is evidently an expensive one. Quite often the WHERE statement takes longer than IF. As a general rule, WHERE runs faster when you subset up to half of the observations. But the results vary with the operating system, the release of the software, and the number of variables per observation. In all the test results below, a simple DATA step subsets the observations, along the lines of:

```
DATA SUBSET;
SET ORIGINAL;
IF I < 10000;
```

All results in these tables illustrate the percentage of CPU time saved by switching to WHERE. (Negative numbers indicate cases in which WHERE took longer than IF.) When ORIGINAL contained 30 variables, these were the percentage savings:

	Operating System			
% Obs Selected	MVS	Win NT	UNIX	OpenVMS
20%	5%	21%	33%	22%
30%	-1%	11%	26%	20%
50%	-17%	-4%	11%	6%
80%	-36%	-15%	-12%	-18%

1. buffer
2. PDV

When ORIGINAL contained 100 variables, these were the percentage savings:

	Operating System			
% Obs Selected	MVS	Win NT	UNIX	OpenVMS
20%	26%	29%	38%	14%
30%	17%	24%	34%	21%
50%	5%	20%	22%	9%
80%	-15%	4%	3%	2%

The 14% and 21% savings under OpenVMS are reported accurately here. Although each test was run three times, there was enough random fluctuation in the results to generate larger savings in the larger subset (30% of the observations).

As the incoming data contain more and more variables, WHERE gains relative to IF. This makes sense because the greater the number of variables, the more WHERE saves by skipping an observation without reading it in. However, this result also implies that good programming technique, applying KEEP= on the SET statement, affects whether WHERE will run faster than IF. Consider these two DATA steps:

```
DATA MILK;
SET COWS;
IF NAME = 'ELSIE';
CUPS = 2*PINTS + 4*QUARTS;
KEEP NAME CUPS;

DATA MILK;
SET COWS (KEEP=NAME PINTS QUARTS);
IF NAME = 'ELSIE';
CUPS = 2*PINTS + 4*QUARTS;
KEEP NAME CUPS;
```

The first DATA step uses sloppy programming technique, reading in all the variables. The second runs faster, by applying KEEP= to the SET statement. Switching from IF to WHERE might speed up the first DATA step, while slowing down the second one! WHERE will save more when the DATA step sloppily reads in all the variables instead of the needed ones. Of course, the best solution is to fix the SET statement first and then select from IF vs. WHERE.

Other factors influence how and when to subset observations. Some of these factors affect efficiency and some relate to ease of programming.

Factor #1: Must all statements execute for all observations?

The subsetting IF and DELETE statements execute wherever they appear in the DATA step. Therefore, these statements should be placed as early as possible. The second DATA step below executes faster than the first, by moving the subsetting IF:

```
DATA OLDMALES;
SET EVERYONE;
HWRATIO = HEIGHT / WEIGHT;
TOTALQ = SUM(OF Q1-Q4);
IF GENDER='M' AND AGE > 85;

DATA OLDMALES;
SET EVERYONE;
IF GENDER='M' AND AGE > 85;
HWRATIO = HEIGHT / WEIGHT;
TOTALQ = SUM(OF Q1-Q4);
```

These DATA steps contain identical statements (in a different order) and produce identical output. However, the first calculates HWRATIO and TOTALQ for each observation in EVERYONE. The subsetting IF then deletes most of those observations. Why should the program calculate HWRATIO and TOTALQ for deleted observations? The second DATA step corrects that process by deleting unwanted observations first and then calculating HWRATIO and TOTALQ for the remaining subset only.

Factor #2: Which subsets will be needed?

When a single procedure uses a single subset, the WHERE statement can eliminate a DATA step. For example, the DATA step in this program is unnecssary:

```
DATA NEWYORK;
SET USA;
WHERE STATE='NY';

PROC MEANS DATA=NEWYORK;
VAR POP;
```

A faster program would subset in the PROC step instead of the DATA step:

```
PROC MEANS DATA=USA;
WHERE STATE='NY';
VAR POP;
```

This new program subsets and analyzes in one step. By reading the data only once, it runs faster.

Note that the second PROC MEANS takes longer than the first! The second program runs faster because it eliminates a DATA step. However, if a subset existed already containing all the NY observations, it would be faster to use the subset. The reason for that stems from the nature of the WHERE statement. While it lets you subset observations in a PROC step, it still must process every observation in the data set. It processes some observations by throwing them out of the analysis, and it processes others by including them in the calculations. Still, the WHERE statement must process every observation.

When a program processes several subsets, multiple WHERE state-ments can be much slower than an extra DATA step. (Similar consider-ations apply when processing the same subset multiple times.) For example, the following program is inefficient:

```
PROC MEANS DATA=USA;
WHERE STATE='NY';
VAR POP;

PROC FREQ DATA=USA;
WHERE STATE='MA';
TABLES GENDER;

PROC MEANS DATA=USA;
WHERE AGE > 65;
VAR POP;
```

This program examines every observation in USA three times. A faster program would use a DATA step to create the three subsets, and then it would feed each subset into a separate procedure. Refer to the next chapter, File Handling, for examples of creating multiple subsets in one DATA step. For now, though, recognize that each DATA or PROC step examines the entire data set. If your program analyzes several subsets, plan on creating all of them in a single DATA step, rather than using WHERE for each procedure.

One final warning here. The WHERE statement requires a significant amount of time to look through a data set and determine which observations should be used. Omit unnecessary WHERE conditions. Compare these procedures, for example:

```
PROC MEANS DATA=USA;
VAR POP;

PROC MEANS DATA=USA;
VAR POP;
WHERE POP > .;
```

Except for the values of _FREQ_ and NMISS, these two PROC MEANS produce identical results. However, the second PROC FREQ takes significantly longer to run. Don't add a WHERE condition unless you really need it!

Factor #3: Will DATA steps require BY or END= variables?

The DATA step below attempts to calculate subtotals for each STATE. However, because IF subsets differently compared to WHERE, switching from one to the other is likely to produce different results:

```
DATA SUBTOTAL;
SET COUNTRY END=E;
BY STATE;
WHERE /* vs. IF */ AMOUNT < 10;
IF FIRST.STATE THEN TOTAL = 0;
TOTAL + AMOUNT;
IF LAST.STATE;
N + 1;
IF E THEN PUT 'THERE WERE ' N 'STATES';
```

Because IF reads in each observation, then deletes those that aren't needed, IF assigns values to BY and END= variables differently than WHERE. Consider this possibility for the incoming data set:

STATE	AMOUNT	<<<<<< WHERE >>>>>			<<<<<<< IF >>>>>>>		
		FIRST.	LAST.	E	FIRST.	LAST.	E
CA	20				1	0	0
CA	5	1	0	0	0	0	0
CA	8	0	1	0	0	0	0
CA	15				0	1	0
NY	30				1	0	0
NY	3	1	1	1	0	0	0
NY	12				0	1	1

The software sets FIRST.STATE to 1 when reading in the first observation for each STATE. However, WHERE reads in just some of the observations while IF reads in all the observations. Therefore, IF can delete an observation that has FIRST.STATE or LAST.STATE equal to 1. The result: after subsetting with IF, the remaining observations may never have FIRST.STATE, LAST.STATE, or E equal to 1. That makes programming much more difficult! Selecting the faster statement becomes irrelevant.

Factor #4: Would a subsetting WHERE be legal?

The WHERE statement faces several restrictions that a subsetting IF does not. The WHERE statement cannot be used

- when reading from raw data.

- when the OBS= or FIRSTOBS= option is in effect (except when they take on their default values of MAX and 1, respectively).

- to reference variables which are not contained in the incoming SAS data set(s).

On the other hand, SAS procedures can use a WHERE statement but they cannot use a subsetting IF.

To circumvent the second limitation, avoid setting the OBS= option globally:

```
OPTIONS OBS=50;
```

Instead, limit observations in an early step that does not use a WHERE statement:

```
DATA NEW;
SET OLD (OBS=50);
TOTAL = 2*PINTS + 4*QUARTS;
```

Let the global OBS option remain at its default value of MAX.

Factor #5: Does a suitable index exist for WHERE processing?

When the software sees a WHERE statement, it automatically:

1. Investigates whether an index exists based on the variable(s) in the WHERE clause (or that uses the WHERE variables as the primary keys).

2.   If an index exists, estimates whether the data would be retrieved faster by using the index.

3.   Uses the index if that would be faster.

Faster may not mean better! There are two possible risks. First, when retrieving observations via an index, the order may not be the same as the sequential order in the incoming data. And second, the software's estimate concerning the fastest method may be inaccurate. Chapter 4 explains how to override the software's choice on whether to use an index.

For Multiple Passes, Subset Variables Early

Because this program processes only two variables, it contains a hidden inefficiency:

```
PROC MEANS DATA=IN.WIDE;
VAR AMOUNT;
CLASS STATE;

PROC CHART DATA=IN.WIDE;
VBAR STATE / SUMVAR=AMOUNT;

PROC FREQ DATA=IN.WIDE;
TABLES STATE*AMOUNT;
FORMAT AMOUNT MYCAT.;
```

For each procedure to read in AMOUNT and STATE, it must retrieve the entire data set IN.WIDE. For a single pass through the data, that approach would run fastest. But for multiple passes, each pass must retrieve many extra variables. Instead, add a DATA step to create a narrower data set:

```
DATA NARROW;
SET IN.WIDE (KEEP=AMOUNT STATE);
```

By operating on the narrow data set, each procedure runs faster. Despite the extra CPU time needed for the DATA step, the procedures can run fast enough to make up the difference. Of course, you will experience variation in the results, depending on how many variables are in the wide data set vs. the narrow one, and how many times your program processes the narrow data set.

Section 3: General Techniques for Reading Data

Store Permanent SAS Data Sets

Most programs create SAS data sets and then analyze those data sets using SAS procedures. If many programs analyze the same data, you can create the permanent SAS data sets once, skipping that step in subsequent programs. Even when the analysis portion of the program uses DATA steps rather than PROC steps, this tip still saves. The DATA step reads SAS data sets faster than it reads raw data.

The benefits are obvious, but the costs can be subtle. Obviously, you will need more permanent storage space. Less obviously, you will need to keep track of these data sets. Which is your master data, the SAS data set or the raw data from which it came? Is the SAS data set based on the most recent version of the raw data? When can you afford to delete the SAS data set or the raw data?

During the testing and debugging stage, some programs can run against a subset or sample of the data. This chapter explores various sampling techniques. But some programs must run against large amounts of data, even in the testing phase. In that case, a program that would normally create temporary SAS data sets might benefit by creating permanent ones. It pays to save the data created by the first half of the program until mistakes are corrected in the second half. Testing runs much faster if you follow these steps:

1. Permanently save SAS data sets created in the first half of the program.

2. Comment out the first half of the program so it won't run during subsequent tests.

3. Once all the bugs are out, change the first half of the program so that it creates temporary data sets only.

Two programming tools make this process practical. First, the USER option lets you save single level data set names in a permanent library. The first half of this program uses single-level names for all data sets:

```
DATA TEMP;   /* Needed for MEANS */
INFILE RAWDATA;
INPUT ID $8. (V1-V50) (2.);

PROC MEANS DATA=TEMP NWAY NOPRINT;
CLASS ID;
VAR V1-V50;
OUTPUT OUT=SMALL MEAN=;
```

To permanently save TEMP and SMALL, you don't have to edit all the code, changing single-level names to two-part names. Assuming the program defines DEBUG as a permanent SAS data library, simply add this statement:

Libruelly

```
OPTIONS USER=DEBUG;
```

From this point forward, the program will automatically save all single level data sets names, in this case TEMP and SMALL, in that library. Once the program is working, remove the OPTIONS statement and delete the SAS data sets DEBUG.TEMP and DEBUG.SMALL, which were created by the testing process.

This overall plan requires a method of commenting out the first half of the program. Let's rule out as totally impractical placing an asterisk at the beginning of each statement. Even using /* and */ may not be practical. Consider what happens if we were to apply that technique to the sample program above:

```
DATA TEMP;   /* Needed for MEANS */
INFILE RAWDATA;
INPUT ID $8. (V1-V50) (2.);

PROC MEANS DATA=TEMP NWAY NOPRINT;
CLASS ID;
VAR V1-V50;
OUTPUT OUT=SMALL MEAN=;
   */
```

This style of comment does not have to be "balanced" like quotes or parentheses. Therefore, this program comments out only the first statement:

```
/*
DATA TEMP;   /* Needed for MEANS */
```

The extra /* in the middle of the program is part of the commented out text, not a set of symbolic characters. The INFILE statement appears without a DATA statement and generates an error message. In addition, the final */ begins a comment statement, commenting out the first statement in the remainder of the program.

To easily comment out a block of code that may contain embedded comments, define the entire block as a macro:

```
%MACRO TESTED;

DATA TEMP;   /* Needed for MEANS */
INFILE RAWDATA;
INPUT ID $8. (V1-V50) (2.);
```

```
PROC MEANS DATA=TEMP NWAY NOPRINT;
CLASS ID;
VAR V1-V50;
OUTPUT OUT=SMALL MEAN=;

%MEND TESTED;
```

True, the program will compile the macro. But the macro never executes, and it is easy to remove the %MACRO and %MEND statements later or to move them as additional portions of the program complete the testing phase. Note that a CARDS statement is illegal inside a macro definition.

Test Programs on a Subset/Sample

The smaller the number of observations, the faster the program runs. So in the developmental stage of writing a program, test against a sample or a subset until you have worked out the bugs. Here is one way to select a sample:

```
OPTIONS OBS=50; /* or 0 or 1000, pick a number */
```

Now the program runs on the first 50 observations from any source of data. If the first 50 observations won't do, you can select 50 from the middle of the data:

```
OPTIONS FIRSTOBS=51 OBS=100;
```

Note that OBS= indicates the last observation selected, not the total number. This OPTIONS statement selects 50 observations in total, starting with the 51st.

Do not leave the FIRSTOBS option in effect throughout a program (except with a value of 1). With the above options in effect, this program would produce no report regardless of the number of observations in OLD:

```
PROC SORT DATA=OLD OUT=NEW;
BY STATE;

PROC PRINT DATA=NEW;
```

Because of FIRSTOBS and OBS, NEW contains only 50 observations (the 51st through 100th from OLD). When these options remain in effect, PROC PRINT prints all of NEW beginning with the 51st observation. Because NEW contains only 50 observations, there is no report. As an alternative, apply FIRSTOBS (possibly OBS as well) as a data set option. Replace the previous program with:

```
PROC SORT DATA=OLD (FIRSTOBS=51 OBS=100) OUT=NEW;
BY STATE;

PROC PRINT DATA=NEW;
```

In addition to printing all of NEW, this version of the program lets you use a WHERE statement later in the program. WHERE would be illegal with FIRSTOBS or OBS in effect.

Taking a block of 50 observations here or there may not give you a sufficient sample for testing. Any of these situations would require a different sample:

- Your program might later subset the observations, based on data values. By applying these subsetting criteria, that original block of 50 might dwindle down to zero observations.

- Your report might require a sample that better reflects the entire data set. When producing a table with PROC FREQ or PROC TABULATE (or even a summary with PROC REPORT), your sample must reflect the data values across your entire data set. Other-wise, the table based on the sample will not resemble the final table based on all the data.

- For statistical applications, a block of 50 observations may be too clustered. You may want to select a more dispersed sample or even a random sample.

Let's briefly examine some programming techniques that handle these situations.

Selecting a Sample When Subsetting

The global option OBS= generates poor results when subsetting:

```
OPTIONS OBS=50;

DATA TEST50;
INFILE RAWDATA;
INPUT AMOUNT 1-8;
IF AMOUNT > 900;
```

The data set TEST50 will contain zero observations, whenever the first 50 AMOUNTs are less than 900. A common beginner's mistake is to adjust the value of the OBS= option, bumping it up to 100 or 1000 or 5000 until the program locates a suitable number of observations. Besides running the program several times and getting no results, this approach may generate another inefficient outcome. The final run may be based on 2000 observations instead of 50 if OBS=5000 locates many large AMOUNTs. A better program would count the sample size in the DATA step:

```
DATA TEST50;
INFILE RAWDATA;
INPUT AMOUNT 1-8;
IF AMOUNT > 900;
N + 1;
OUTPUT;
IF N=50 THEN STOP;
```

Now you are guaranteed to run on exactly 50 observations (unless the data actually contain fewer than 50 large AMOUNTs), and the DATA step never reads unneeded observations. In addition, the OBS= option is not in effect, so later DATA and PROC steps can use a WHERE statement. Notice that the DATA step requires an OUTPUT statement. Without it, the STOP statement would have halted the DATA step before outputting the 50th observation. The sample would have contained 49 observations instead.

Sampling Every Nth Observation

A block of 50 observations may fail to mirror the larger data set. To select more diverse test data, one sampling approach is to select every Nth observation. When reading raw data, the software can easily select every 25th observation, beginning with the 20th:

```
DATA FOUR_PCT;
INFILE RAWDATA;
INPUT #20 AMOUNT 1-8 #25;
```

If the number of raw data lines is not a multiple of 25, you will get a harmless message about LOST CARD. Similarly, this DATA step selects the same set of observations from a SAS data set:

```
DATA FOUR_PCT;
SET LARGE.FILE;
IF MOD(_N_, 25)=20;
```

The MOD function divides the first number by the second and returns the remainder. We could trade off here, writing a more complex program but one that reads fewer observations from LARGE.FILE:

```
DATA FOUR_PCT;
DO I=20 TO TOTALOBS BY 25;
    SET LARGE.FILE POINT=I NOBS=TOTALOBS;
    OUTPUT;
END;
STOP;
```

The NOBS= option creates TOTALOBS as the software compiles the SET statement. Automatically, TOTALOBS holds the number of observations in LARGE.FILE. The POINT= option forces the SET statement to retrieve selected observations (based on the value of I) instead of reading the observations sequentially. While the POINT= option slows down the SET statement, the new DATA step executes the SET statement for only 4% of the observations. Therefore, the new program runs about 75% faster.

Selecting a Random Sample

Selecting every 25th observation is different from selecting a random sample. Sometimes, for statistical purposes, the program must select the observations randomly. If the program truly requires a random sample, one approach is:

1. Assign a random number to each observation.

2. Select those observations that possess the lowest random numbers.

Unfortunately, this approach requires a multi-step process, such as:

```
DATA ASSIGNED;
SET ORIGINAL;
SORTVAR=RANUNI(0);

PROC SORT DATA=ORIGINAL;
BY SORTVAR;

PROC MEANS DATA=ASSIGNED (OBS=200);
VAR AMOUNT;
```

The RANUNI function generates a random number between zero and one, which is evenly distributed within that interval. Therefore, the observations with the 200 smallest random numbers make up a random sample of size 200.

If the objective is to select a random sample of exactly 200 observations, this type of approach is necessary. However, if you can relax the requirements slightly, you can get away with one step instead of two. For example, if the sample size should be approximately 200 (but does not have to be exactly 200), you can once again utilize the NOBS= option on the SET statement:

```
DATA _200_;
SET LARGE.FILE NOBS=TOTALOBS;
IF RANUNI(0) < 200/TOTALOBS;
```

If the sampling methodology permits selecting the same observation more than once, you can randomly select exactly 200 observations in one step. Again, the NOBS= option in the SET statement retrieves the total number of observations in LARGE.FILE. And the POINT= option on the SET statement again retrieves specific observations instead of reading all observations sequentially. The program becomes:

```
DATA _200_;
DO I=1 TO 200;
    OBSNO = CEIL(TOTALOBS * RANUNI(0));
    SET LARGE.FILE POINT=OBSNO NOBS=TOTALOBS;
    OUTPUT;
END;
STOP;
```

The CEIL function operates on an evenly distributed number greater than zero and less than the total number of observations in the data set. Therefore, OBSNO is an evenly distributed integer, with a minimum value of 1 and a maximum value of the total number of observations in the data set. As noted, these statements can select the same observation twice.

Finally, a one-step approach can select an exact sample size in a *nearly* random fashion. For example:

```
DATA _200_;
RETAIN NEEDED 200 TOTALOBS;
IF NEEDED=0 THEN STOP;
SET LARGE.FILE NOBS=TOTALOBS;
IF _N_ > 1 THEN TOTALOBS = TOTALOBS - 1;
IF RANUNI(0) < NEEDED/TOTALOBS;
NEEDED = NEEDED - 1;
```

This DATA step uses TOTALOBS and NEEDED to track the number of observations remaining in LARGE.FILE and the number still needed to complete a sample of 200 observations. It incorporates some key features:

- The NOBS= option in the SET statement creates TOTALOBS once at compile time, not once per observation. Therefore, the DATA step can modify TOTALOBS as it reads observations from LARGE.FILE.

- The subsetting IF deletes observations when RANUNI generates too large a random number. When the randon number is small enough, however, the program subtracts one from NEEDED and outputs the observation.

This does not constitute a random sample by traditional statistical standards. In a random sample, each observation must have the same probability of selection as all others. In this program, that probability changes from observation to observation. As the DATA step begins, the probabilities are roughly equal. After all, 200/10000 is about the same as 200/9999. However, this slight difference becomes more pronounced as the DATA step continues. If we were to reach the end of the data set, the final value of NEEDED would be 1 and the final value of TOTALOBS would be 1, yielding a 100% chance of selecting the final observation.

Can You Still Beat the DATA Step Loop?

Under Version 5 of the SAS software, a simple DATA step such as this one involved two programs:

```
DATA TOTALS;
SET COWS;
CUPS = 2*PINTS + 4*QUARTS;
```

One program (the Supervisor) called a second program to execute the SET and assignment statements. When the second program finished, the Supervisor would take over again, outputting the observation and resetting CUPS to missing. Then the software would go through the same process for the next observation. This process of one program calling another was relatively expensive (up to 15% of the cost of the DATA step). To eliminate that work, a Version 5 DATA step could have read COWS in a DO loop:

```
DATA TOTALS;
DO UNTIL (NOMORE);
   SET COWS END=NOMORE;
   CUPS = 2*PINTS + 4*QUARTS;
   OUTPUT;
END;
STOP;
```

Version 6 of the SAS System eliminated most of the efficiencies of switching to a DO loop. In a Version 6 DATA step, the Supervisor lets the second program run through the entire data set (unless the program encounters an error condition). However, all DATA steps still perform some work at the DATA statement. At that point, the software must still:

- generate appropriate messages if _ERROR_=1.

- determine whether the current observation should be output.

- after outputting, run through all variables that are not retained, setting them back to a missing value.

- reset _ERROR_ and _N_.

Although the DO loop can eliminate some of that work, just adding the DO loop increases the workload. By switching to a DO loop, sample tests generated a maximum savings of 3% of the DATA step's CPU time on any operating system. In some cases, the original program ran faster. Chapter 10 presents the details. Among these inconsistencies, one small opportunity for savings emerges. You can speed up your DATA steps by retaining all variables (as long as the logic still generates the right result). The software now has less work to perform between observations. Note that the software automatically retains all variables read from SAS data sets. So this step applies to variables input from raw data as well as variables created by DATA step programming statements.

A Small Workshop

Modify the programs below to reduce the CPU time. Focus on reading data, but also consider other factors such as time spent on data manipulation.

Consider each program in two stages. First, note all the unnecessary use of CPU time. Then fix the program to reduce the CPU time.

Program #1

A side note: the INTNX function returns the first day of a time period. Thus if CURRENT is any date during April 1999, the dates represented by START and FINISH would be January 1, 1999 and March 31, 1999.

```
DATA NEW (KEEP=NAME AGE HEIGHT WEIGHT DATE
                SLIMNESS START FINISH);
SET OLD;
SLIMNESS = HEIGHT/WEIGHT;
CURRENT = TODAY();
START = INTNX('MONTH', CURRENT, -3);
FINISH = INTNX('MONTH', CURRENT, 0) - 1;
IF START <= DATE <= FINISH;
```

Program #2

In this program, assume that all character values begin with a nonblank character.

```
DATA INFANTS;
INFILE CHILDREN;
INPUT NAME $ 1-20   AGE 21-23   HEIGHT 24-25
      WEIGHT 26-28  SEX $ 29   ADDRESS $ 30-49
      CITY $ 50-69  STATE $ 70-71   ZIP 72-76;
IF AGE < 3;
```

Program #3

Assume that the report from PROC MEANS is the only output needed. The colon modifier compares character strings based on the length of the shorter string. Chapter 7 investigates this tool in more detail.

```
DATA MEMBERS;
INFILE PROSPECT;
INPUT ID $ 1-9  AGE 10-12   GENDER $ 13
      LASTNAME $ 14-28   FNAME $ 29-40;

PROC MEANS DATA=MEMBERS MEAN N;
VAR AGE;
WHERE GENDER='M' AND LASTNAME =: 'JO';
```

Workshop Solutions

Program #1, the extra work:

- The SET statement reads all variables instead of the needed ones only.

- Because CURRENT, START, and FINISH remain constant, the program should compute those once and retain their values.

- The program computes SLIMNESS for each observation, some of which get deleted later by the subsetting IF.

A faster program would be:

```
DATA NEW (DROP=CURRENT);
SET OLD  (KEEP=NAME AGE HEIGHT WEIGHT DATE);
IF _N_=1 THEN DO;
   CURRENT = TODAY();
   START = INTNX('MONTH', CURRENT, -3);
   FINISH = INTNX('MONTH', CURRENT, 0) - 1;
END;
RETAIN START FINISH CURRENT;
IF START <= DATE <= FINISH;
SLIMNESS = HEIGHT/WEIGHT;
```

Program #2, the extra work:

The program reads in:

- all variables, not just AGE, before the subsetting IF.

- ZIP as numeric instead of character.

- character variables without a $CHAR informat.

- consecutive fields yet still specifies starting columns.

A faster program would be:

```
DATA INFANTS;
INFILE CHILDREN;
INPUT AGE 21-23 @;
IF AGE < 3;
INPUT @1 NAME $CHAR20.  @24 HEIGHT 2.
      WEIGHT 3.  SEX $CHAR1.  ADDRESS $CHAR20.
      CITY $CHAR20.  STATE $CHAR2.  ZIP $CHAR5.;
```

Program #3, the DATA step:

- reads in extra variables (ID and FNAME) which never get used.

- reads all of LASTNAME instead of the first two characters.

- keeps all observations in MEMBERS, so that PROC MEANS must later determine which observations to omit and which to process.

- reads AGE for every observation instead of a subset.

A faster program would be:

```
DATA MEMBERS;
INFILE PROSPECT;
INPUT @13 GENDER $CHAR1. LASTNAME $CHAR2. @;
IF GENDER='M' AND LASTNAME='JO';
INPUT AGE 10-12;
KEEP AGE;

PROC MEANS DATA=MEMBERS MEAN N;
VAR AGE;
```

Finally, if MEMBERS were never needed again, you could eliminate PROC MEANS entirely. The DATA step is quite capable of calculating the MEAN and N statistics:

```
DATA _NULL_;
IF NOMORE THEN DO;
   MEAN = TOTAGES/N;
   FILE PRINT;
```

```
     PUT 'Mean age of all ' N
         'members is ' MEAN;
END;
INFILE PROSPECT END=NOMORE;
INPUT @13 GENDER $CHAR1. LASTNAME $CHAR2. @;
IF GENDER='M' AND LASTNAME='JO';
INPUT AGE 10-12;
IF AGE > .;
TOTAGES + AGE;
N + 1;
```

The check for end of file must appear at the top of the DATA step. If it were to appear later, one of the subsetting IFs might delete the final observation. In that case, the DATA step would fail to produce a report.

CHAPTER 3

Reporting

Fast reports or slow reports? Sometimes the choice is yours.

Specify Formats

Many procedures spend CPU time to format each page of output. For example, PROC FREQ adjusts the report structure based on the number of rows and columns in the table. PROC PRINT investigates the length of data values on the page to determine the appropriate column width for printing each variable. Especially in the case of PROC PRINT, you can save the software some of the formatting decisions by specifying formats for variables. PROC PRINT tests cut out two thirds of the CPU time by formatting every variable.

PROC PRINT output takes on a uniform appearance when you format all the variables being printed. You don't need to specify the UNIFORM option, which would force the software to look through the data twice (once to find the longest values and then a second time to print them based on the maximum length of each variable). By formatting the variables, you get the benefit of a uniform appearance (based on the widths in the FORMAT statement) without the drawback of forcing the SAS System to look through the data twice.

Use _NULL_ Data Sets

When you don't need an output data set, use _NULL_ as the name in the DATA statement. _NULL_ is a keyword as a data set name. It actually means "Don't create a SAS data set. We are going through the motions of a DATA step for some other purpose, probably to generate a report with PUT statements." So the third DATA statement below runs faster than the first two:

```
DATA TEMP;
DATA;
DATA _NULL_;
```

The first takes CPU time to output observations into TEMP. The second also takes CPU time to output observations. However, the program is letting the software select the name for that output data set. Only the third creates no output data set at all. Obviously, the storage space requirements are smaller when there is no output data set.

Although most DATA _NULL_ steps generate reports, some write raw data. For those that also read raw data, the next two techniques may come in handy.

Use PUT _INFILE_

Each time an INPUT statement executes, it changes the value of the
automatic variable _INFILE_. In these two DATA steps, only the second
one utilizes _INFILE_:

```
DATA _NULL_;
INFILE RAWDATA;
INPUT LINE $CHAR100.;
IF RANUNI(0) < 0.2;
FILE OUTFILE;
PUT LINE $CHAR100.;

DATA _NULL_;
INFILE RAWDATA;
INPUT;
IF RANUNI(0) < 0.2;
FILE OUTFILE;
PUT _INFILE_;
```

Both INPUT statements create the same values for _INFILE_. However,
the first DATA step performs extra work by creating the variable LINE. In
addition, the second DATA step incorporates these benefits:

- In Version 6 and earlier releases, character variables have a
  maximum length of 200 characters. If the line length were
  longer, the top DATA step would need to create additional
  character variables. _INFILE_ can be tens of thousands of
  characters long.

- Writing _INFILE_ is marginally slower than writing one variable,
  but is much faster than writing a large set of variables. In practical applications, the variable LINE may actually represent a
  much larger set of variables.

If the output file requires just the first portion of the original raw data line, the _INFILE_ approach still works. Add the LENGTH= option:

```
DATA _NULL_;
INFILE RAWDATA LENGTH=L;
INPUT;
IF RANUNI(0) < 0.2;
L=100;
FILE OUTFILE;
PUT _INFILE_;
```

[handwritten annotation]

By resetting L to 100, the program writes out just the first 100 characters of the raw data lines.

Use the SHAREBUFFERS Option

When writing from one raw data file to another, SAS software uses one buffer to hold the incoming raw data lines and a second buffer to hold the written lines. If the incoming and outgoing files are very similar, you can use one storage area and avoid copying from the input to the output buffer. Consider this program, for example:

```
DATA _NULL_;
INFILE RAWDATA;
INPUT @4 (V1-V50) (1.);
TOTSCORE = SUM(OF V1-V50);
FILE OUTFILE;
PUT _INFILE_ @101 TOTSCORE 3.;
```

The output file contains a copy of the input file, with TOTSCORE in columns 101 to 103, so the input and output files can share the same raw data buffer:

```
DATA _NULL_;
INFILE RAWDATA SHAREBUFFERS;
INPUT @4 (V1-V50) (1.);
TOTSCORE = SUM(OF V1-V50);
PUT _INFILE_ @101 TOTSCORE 3.;
```

Summarize Data Before Running PROC TABULATE

PROC TABULATE provides excellent reporting features: combining variables in a dimension of the table as well as printing statistics (not merely frequency counts) in the cells of the table. However, PROC SUMMARY calculates nearly the same statistics and, under Version 5, would calculate them faster than PROC TABULATE. This used to be a rare case where two procedures ran faster than one:

1. Summarize the data using PROC SUMMARY.

2. Use the PROC SUMMARY output data set as the input to PROC TABULATE.

Version 6 rewrote the internal code of PROC TABULATE, speeding up the calculations. Although the two-step approach now takes longer, it still has other benefits:

-   You can process the output data set further, such as in a DATA step, until it contains exactly the right information needed by PROC TABULATE.

-   You can experiment with the format of the table (sometimes necessary with PROC TABULATE) until it takes on exactly the right appearance. Experiments using small summary data sets run relatively quickly.

CHAPTER 4

File Handling

Overview

Chapter 2 examined how to read data, focusing on reading raw data and basic techniques for reading SAS data. This chapter goes a step farther, introducing PROC steps, fancier DATA steps, and combinations of the two.

Typically, each DATA and PROC step reads in a SAS data set. Cut back on the number of DATA and PROC steps, and you will cut back on the corresponding CPU time. Cutting back may involve eliminating poor habits, or it may involve expanding your repertoire to include

---

Legend for Icons

CPU time or storage space
Black stars = technique applies frequently
White stars = technique applies infrequently
3 black or white stars - maximum savings
2 black or white stars - medium savings
1 black or white star - smallest savings

Fluctuating savings
Savings can range from large to nonexistent to negative, depending on your operating system or data characteristics.

Archaic
No savings under current release.

more sophisticated programming techniques. This chapter breaks those topics down into three sections:

1. Eliminating poor programming habits.

2. Altering a data set without reading in all the data values.

3. Miscellaneous file handling techniques.

Section 1: Eliminate Poor Habits

In this context, "poor habits" means processing a data set an extra time, without good reason. Most bad habits are easy to correct, as these examples illustrate.

PROC FREQ Works on Unsorted Data

You can sort your data before running PROC FREQ, but it won't make the FREQ procedure run any faster. Sorting is totally unnecessary in this program:

```
PROC SORT DATA=RAINBOW;
BY COLOR;

PROC FREQ DATA=RAINBOW;
TABLES COLOR;
```

Simply eliminate the PROC SORT and save a large portion of the CPU time. When PROC FREQ doesn't include a BY statement, it doesn't need sorted data. Chapter 5 (Sorting Data) illustrates PROC FREQ with a BY statement, which can be used as a method to trade off resources, adding CPU time of PROC SORT to make up for memory limitations of PROC FREQ.

Don't Sort by Extra Variables

The PROC CHART below requires sorted data because it contains a BY statement:

```
PROC SORT DATA=HOTELS;
BY LOCATION AGE;

PROC CHART DATA=HOTELS;
BY LOCATION;
HBAR AGE / DISCRETE;
```

However, PROC SORT is still performing unnecessary work. Sorting BY LOCATION will order the data sufficiently for PROC CHART. Sorting by AGE within each LOCATION is wasted work.

Process Permanent SAS Data Sets

When we begin to write SAS programs, we learn how our programs first create a SAS data set and then analyze the data set with SAS procedures. Apply this process to your programs as a whole, not to each program individually. When an earlier program permanently saves a SAS data set, later programs can begin with PROC steps. This first example contains an extra DATA step:

```
DATA TEMP;
SET SAVED.SASDATA;

PROC MEANS DATA=TEMP;
VAR HEIGHT WEIGHT AGE;
```

PROC MEANS can easily work on the permanently saved data:

```
PROC MEANS DATA=SAVED.SASDATA;
VAR HEIGHT WEIGHT AGE;
```

In similar fashion, this program contains an extra DATA step:

```
DATA TEMP;
SET SAVED.SASDATA;

PROC SORT DATA=TEMP;
BY ID;
```

Even when you don't want to sort the permanent SAS data set, you can use OUT= to eliminate the DATA step:

```
PROC SORT DATA=SAVED.SASDATA OUT=TEMP;
BY ID;
```

Perform All Data Manipulation in a Single DATA Step

For purposes of simplicity, this sample program contains only one extra DATA step:

```
DATA NEWYORK;
SET ENTIRE.USA;
IF STATE='NY';

DATA ADDVARS;
SET NEWYORK;
HWRATIO = HEIGHT/WEIGHT;
YEARS = DAYS/365.25;
```

Obviously, the program can perform all the data manipulation in one step:

```
DATA ADDVARS;
SET ENTIRE.USA;
IF STATE='NY';
HWRATIO = HEIGHT/WEIGHT;
YEARS = DAYS/365.25;
```

The data manipulation runs relatively quickly. Reading in and outputting the observations is the time-consuming portion of the DATA step. Do it once instead of twice.

Section 2: Altering a Data Set without Reading All the Data Values

Once you have eliminated bad habits, you can still speed up your file handling. In this section, we consider situations that don't have to read all the data values to accomplish the program's objectives.

Copy with PROC COPY

PROC COPY (as well as the COPY statement within PROC DATASETS) copies SAS data directly from an input to an output destination.

```
PROC COPY IN=EXISTING OUT=NEW;
SELECT DATASET NAMES;
```

This program copies two SAS data sets (DATASET and NAMES) from the EXISTING library to the NEW library. Two DATA steps could perform the job: \FOLDER

```
DATA NEW.DATASET;
SET EXISTING.DATASET;

DATA NEW.NAMES;
SET EXISTING.NAMES;
```

The DATA steps perform more work than PROC COPY by moving observations to intermediate storage locations along the way. These additional storage locations support other capabilities of the DATA step, such as calculating new variables. When your sole purpose is to copy selected data sets, however, PROC COPY works faster. In addition, PROC COPY automatically copies indices from the existing library. The DATA step must instead build an index to the new data set.

Append with PROC APPEND

PROC APPEND (as well as the APPEND statement within PROC DATASETS) copies a SAS data set to the end of another existing SAS data set:

```
PROC APPEND DATA=SMALL BASE=LARGE.DATASET   /* FORCE */;
```

This procedure avoids having to read and write observations from LARGE.DATASET. The DATA step, in contrast, would read in all the data from both sources:

```
DATA LARGER.DATASET;
SET LARGE.DATASET SMALL;
```

By default, PROC APPEND assumes that both data sets have the same structure. However, if the two data sets contain different variables or have different lengths for the same variables, the DATA step would work while PROC APPEND would fail. To force PROC APPEND to combine data sets with different structures, add the FORCE option. While FORCE allows PROC APPEND to combine data sets with different structures, the results will differ from the DATA step results. PROC APPEND drops variables that appear in the DATA= data set but not in the BASE= data set. Of course, if one data set defines a variable as character and the other defines the same variable as numeric, neither method will combine the data sets.

Modify Data Set Structure with PROC DATASETS

The SAS System stores information about each SAS data set separately from the data values themselves. PROC CONTENTS can access this descriptor information and thus display information about the variables in a SAS data set without having to actually read the data values. The software can modify the descriptor information without a DATA step for many types of structural changes such as changing a data set name, changing a variable name, retrieving the number of observations in the data set, or assigning labels, formats, or informats to variables. The slower program would use a DATA step:

```
DATA NEWNAME;
SET OLDNAME;
RENAME LASTNAME=LNAME;
LABEL FNAME='First Name';

PROC DATASETS LIBRARY=WORK;
DELETE OLDNAME;
```
≥
.

This program copies data from OLDNAME to NEWNAME, renaming one variable and labeling another along the way. Once all the work is done, the program deletes the original data set. Because all of the changes involve the structure of the data set, reading in and writing out each observation in OLDNAME is really extra work. PROC DATASETS could make those same changes without reading and writing the data:

```
PROC DATASETS LIBRARY=WORK;
CHANGE OLDNAME=NEWNAME;
MODIFY NEWNAME;
RENAME LASTNAME=LNAME;
LABEL FNAME='First Name';
```
obe

Because PROC DATASETS does not read in the data, this program uses virtually no CPU time, regardless of the size of OLDNAME.

Section 3: Miscellaneous File Handling Techniques

This final section examines assorted file handling situations where careful programming can reduce CPU time.

Repeating a Single Observation

This first example involves a very specific situation. The SAS data set ONE_OBS contains a single observation and 50 variables, while MANY_OBS contains many observations.

The program must duplicate all 50 variables from ONE_OBS on each of the observations from MANY_OBS. The slow method reads both data sets twice:

```
DATA ONE_OBS;
SET ONE_OBS;
MERGEVAR='A';

DATA MANY_OBS;
SET MANY_OBS;
MERGEVAR='A';

DATA BOTH;
MERGE ONE_OBS MANY_OBS;
BY MERGEVAR;
DROP MERGEVAR;
```

The shorter method utilizes a key piece of information: any variable read from a SAS data set is automatically retained. That knowledge would let you shorten the program to a single step:

```
DATA BOTH;
IF _N_=1 THEN SET ONE_OBS;
SET MANYOBS;
```

A few key steps take place as this DATA step executes:

1. As the software leaves the DATA statement for the first time, _N_=1. The software therefore reads the first (and only) observation from ONE_OBS.

2. The second SET statement reads in the first observation from MANY_OBS.

3. At the end of the programming statements, the software returns to the DATA statement and outputs the current observation.

4. The program continues by leaving the DATA statement a second time. As this step takes place, two important features come into play. First, _N_=2 because this is the second instance of leaving the DATA statement. Second, any variables read from SAS data sets are automatically retained. That feature

applies to all the variables because all variables are read using SET statements.

5. The comparison IF _N_=1 is now false, and the first SET statement gets skipped. However, the variable values from ONE_OBS are still retained in the Program Data Vector.

6. The second SET statement now reads the second observation from MANY_OBS, replacing the retained data values from the first observation. Now the program has stored in the Program Data Vector the second observation from MANY_OBS as well as the retained values from ONE_OBS.

7. As usual, the software returns to the DATA statement and outputs the current observation.

Steps 4 through 7 continue until the DATA step has processed every observation in MANY_OBS.

One DATA Step, Multiple Data Sets

A single DATA step can create and combine multiple data sets. This program fails to exploit the DATA step's capabilities when it reads in EVERYONE three times:

```
DATA MALES;
SET EVERYONE;
IF GENDER='M';

DATA FEMALES;
SET EVERYONE;
IF GENDER='F';

DATA WHOKNOWS;
SET EVERYONE;
IF GENDER NOT IN ('F', 'M');
```

One read would have been enough:

```
DATA MALES FEMALES WHOKNOWS;
SET EVERYONE;
IF GENDER='M' THEN OUTPUT MALES;
ELSE IF GENDER='F' THEN OUTPUT FEMALES;
ELSE OUTPUT WHOKNOWS;
```

Besides running faster, this DATA step clarifies the outcome: it clearly handles all possible values of GENDER.

Halt Execution Early

Halt your program when it produces bad results. In this example, the program should encounter good data values for GENDER. The ABORT statement halts the program when the data contain any bad values:

```
DATA MALES FEMALES;
SET EVERYONE;
IF GENDER='M' THEN OUTPUT MALES;
ELSE IF GENDER='F' THEN OUTPUT FEMALES;
ELSE DO;
   PUT _ALL_;
   ABORT;
END;
```

On the other hand, aborting the program wastes any valid work already completed. The program has already output good values of GENDER. Instead, the program might output all data to permanent data sets, then halt the remainder of the program if any bad data exist. For example:

```
DATA OUT.MALES OUT.FEMALES OUT.BADDATA;
SET EVERYONE;
IF GENDER='M' THEN OUTPUT OUT.MALES;
ELSE IF GENDER='F' THEN OUTPUT OUT.FEMALES;
ELSE OUTPUT OUT.BADDATA;
```

```
DATA _NULL_;
IF EOF=0 THEN CALL EXECUTE ('ENDSAS;');
STOP;
SET OUT.BADDATA END=EOF;
RUN;
```

When OUT.BADDATA contains zero observations, EOF has an initial value of 1 and the DATA step generates an ENDSAS statement. (CALL EXECUTE stacks up statements to run once the current DATA step finishes.) Notice how CALL EXECUTE allows you to conditionally generate SAS code outside of a macro definition because it does not use macro %IF %THEN statements.

Not only does the program save all good data permanently, but it executes subsequent steps if all the data are good. On the other hand, any bad records get saved separately and can be fixed and appended to the good data without having to reread EVERYONE.

Here is a slightly more complex example. The data set MASTER should contain no duplicate IDs, and it needs to be sorted. This sample code halts the program if there are any duplicate IDs:

```
DATA _NULL_;
CALL SYMPUT ('BEFORE', PUT(N_OBS, 9.));
STOP;
SET MASTER NOBS=N_OBS;
RUN;

PROC SORT DATA=MASTER NODUPKEY;
BY ID;

DATA _NULL_;
IF AFTER NE &BEFORE THEN CALL EXECUTE('ENDSAS;');
STOP;
SET MASTER NOBS=AFTER;
RUN;
```

Note that the ENDSAS statement cannot be conditionally executed as a DATA step statement. It must be conditionally generated by macro language. If the following statement were to appear in a DATA step, it would always end the program even though the condition is false:

```
IF 0=1 THEN ENDSAS;
```

PROC SQL vs. the DATA Step

PROC SQL can replace most DATA steps that use SET or MERGE statements. This book will not provide definitive answers as to which is faster. There are just too many variations to consider. For details on some additional situations not covered below, refer to the SAS Institute publication *Advanced SAS Programming Techniques and Efficiencies Course Notes.* Instead, we'll compare some basic DATA steps to the matching PROC SQL code. In these examples, the SAS data sets _030VARS and _100VARS form a many to one match, with some mismatches. This preliminary program creates the two data sets:

```
DATA _030VARS;
DO VAR1=50000 TO 1 BY -1;
   OUTPUT;       /* 50,000 obs, 30 variables */
END;
RETAIN VAR2-VAR30 0;

DATA _100VARS;
DO VAR102=1 TO 10;
   DO VAR1=60000 TO 10001 BY -1;
      OUTPUT;   /* 500,000 obs, 100 variables */
   END;
END;
RETAIN VAR103-VAR200 0;
```

The first three test cases compare straightforward DATA steps with the equivalent SQL code. In these tests, the DATA step almost always completed faster than PROC SQL. The first test reads and outputs an entire data set:

```
DATA TEMP;
SET _100VARS;

PROC SQL;
CREATE TABLE TEMP AS SELECT * FROM _100VARS;
```

The DATA step ran 22% faster under MVS and OpenVMS, but PROC SQL ran just as quickly under Windows NT and UNIX. Chapter 10 reports the results of all test programs.

When subsetting variables, the DATA step ran roughly 20% faster:

```
DATA TEMP;
SET _100VARS (KEEP=VAR1 VAR2A VAR3A);

PROC SQL;
CREATE TABLE TEMP AS
SELECT VAR1, VAR2A, VAR3A FROM _100VARS;
```

When subsetting observations with WHERE, both steps ran equally fast:

```
DATA TEMP;
SET _100VARS;
WHERE VAR100='NewValue';

PROC SQL;
CREATE TABLE TEMP AS
SELECT * FROM _100VARS WHERE VAR100='NewValue';
```

In more complex situations, involving a MERGE or JOIN, test results varied with the operating system. In general, the DATA step ran faster even with the cost of sorting added in. While the programs below represent a FULL JOIN only, the same results held for LEFT JOINs and RIGHT JOINs.

```
PROC SORT DATA=_030VARS;
BY VAR1;

PROC SORT DATA=_100VARS;
BY VAR1;

DATA TEMP;
MERGE _030VARS
      _100VARS
BY VAR1;

PROC SQL;
CREATE TABLE TEMP AS
SELECT * FROM _030VARS FULL JOIN _100VARS
ON _030VARS.VAR1=_100VARS.VAR1;
```

The DATA step requires sorted (or indexed) data whereas PROC SQL runs on unsorted data. However, PROC SQL takes much longer when it works on unsorted data. It is actually faster to sort the data sets first and then run PROC SQL (at least in the test case above), rather than letting PROC SQL run against unsorted data.

One case where PROC SQL routinely beats the alternative is when it replaces combinations of sorting, extracting, and printing in a specific order. A SORT and MERGE approach might proceed:

```
PROC SORT DATA=CUSTOMER;
BY CUSTID;

PROC SORT DATA=PAYMENTS;
BY CUSTID;

DATA COMBINED;
MERGE CUSTOMER (IN=IN1)
      PAYMENTS (IN=IN2);
BY CUSTID;
IF IN1 AND IN2;

PROC SORT DATA=COMBINED;
BY YEAR;

PROC PRINT DATA=COMBINED;
```

The matching code in PROC SQL is shorter and clearer, runs faster, and generates a report without creating a data set to hold the combined files:

```
PROC SQL;
SELECT * FROM CUSTOMER, PAYMENTS
WHERE CUSTOMER.CUSTID = PAYMENTS.CUSTID
ORDER BY YEAR;
```

Creating an Index

Creating an index is a relatively straightforward task. The program below creates an index based on the variable STATE, from the data set PERM.SALES:

```
PROC DATASETS LIBRARY=PERM;
MODIFY SALES;
INDEX CREATE STATE;
```

Indices let the software access small amounts of data quickly. The software would use a suitable index in all four situations below:

```
WHERE STATE='NC';

BY STATE;   /* working on unsorted data */

SET LOOKUP KEY=STATE;

MODIFY PERM.SALES LOOKUP;
BY STATE;
```

To be "suitable," an index must meet two conditions. First, it must exist. In the sample statements above, an index with STATE as the primary key must exist, and it must have been created without using the NOMISS option. Second, the software must determine that the index is the fastest method (or the only method) of properly accessing the needed observation(s).

How does the software evaluate whether the index would be faster or if a sequential read of the data would be faster? Under Version 6, the software examines the first page of data and estimates the number of I/Os needed to read the data via the index. (Because the size of the each data set is known, the number of I/Os for a sequential read is a trivial calculation.) This means that the distribution of variables in the WHERE statement within the first page of data is a critical factor for the software to select the faster method. If that distribution is not representative of the distribution across the entire data set, the software can make the wrong choice. Under Version 7, the software stores information within the index about the distribution of each index variable. The software can therefore make much more accurate choices about when to use the index and when to read the data sequentially.

You do have some control over whether the index will be used. Under Version 6, the $IWEIGHT option influences the software's choice of methods. For example:

```
OPTIONS $IWEIGHT=1;    /* favors using the index */
OPTIONS $IWEIGHT=100;  /* neutral weighting      */
OPTIONS $IWEIGHT=500;  /* favors sequential read */
```

The value of $IWEIGHT is the percentage factor applied to an indexed read. For example, if the software estimates 9 I/Os for a sequential read but 5 I/Os for an indexed read, it computes the cost of a sequential read as 9 * 100 and the cost of an indexed read as 5 * $IWEIGHT. Based on these formulas, the software selects the lower cost method.

Version 7 adds two new options:

```
SET OLD (IDXWHERE=YES);        /* use the best index   */
SET OLD (IDXWHERE=NO);         /* ignore all indices   */
SET OLD (IDXNAME=index name);  /* must use this index */
```

For additional information, refer to a paper by Steve Beatrous and Billy Clifford entitled "Sometimes You Get What You Want: SAS I/O Enhancements for Version 7," published in *Proceedings of the Twenty-Third Annual SAS Users Group International Conference*.

As a general rule, indices that read a small percentage of the observations speed up programs. Indices that read all the data dramatically slow down programs. Assuming that an index based on STATE exists, consider these two PROC PRINTs:

```
PROC PRINT DATA=USA;
VAR CITY POP;
WHERE STATE='NC';

PROC PRINT DATA=USA;
VAR CITY POP;
BY STATE;
```

In the first PROC PRINT, the software would estimate whether it could retrieve the NC observations faster by using the index versus by reading sequentially through the entire data set. The software calculates whether it should use the index.

In the second PROC PRINT, the software would examine whether USA is sorted by STATE. If it is not, the software would retrieve the data in sorted order by using the index. In this case, using the index can vastly increase the CPU time. SAS software does not retrieve observations directly from the incoming SAS data set to the Program Data Vector. Instead, it brings in a page of observations and then reads an observation from that page into the Program Data Vector. When using an index, your program will frequently switch pages to locate the next observation. The inefficiency can be staggering. For a related article, refer to Ian Whitlock's paper, "Why Did This Code Take 22 Hours to Run?", published in *Proceedings of the Tenth Annual NorthEast SAS Users Group Conference*, pages 347-348.

A Small Workshop

Problem #1

Reduce the CPU time in the program below, assuming that PINTS and QUARTS do not exist in the original data set.

```
DATA NEWLIB.COW;
SET OLDLIB.COW;
PINTS=PTS;
QUARTS=QTS;
DROP PTS QTS;
```

Problem #2

A high school keeps detention records in a SAS data set called DETENT. There is one observation per detention, including the key variables NAME ($ 20) and DATE (a SAS date).

Today's list of detentions arrives as raw data, with NAME in columns 1 through 20.

Create a SAS data set HISTORY by extracting from DETENT all observations that match a NAME on today's list. Solve this problem in two ways: (1) a standard SORT and MERGE approach, and (2) creating a format based on today's list. If a student never received detention before today, HISTORY should contain no observations for that student.

Workshop Solutions

Problem #1

The essential steps are copying data and renaming variables. Neither step requires reading data values into the Program Data Vector:

```
PROC COPY IN=OLDLIB OUT=NEWLIB;
SELECT COW;

PROC DATASETS LIBRARY=NEWLIB;
MODIFY COW;
RENAME PTS=PINTS QTS=QUARTS;
```

Problem #2

The SORT and MERGE approach requires sorting and merging both data sets:

```
PROC SORT DATA=DETENT;
BY NAME;

DATA TODAY;
INFILE RAWDATA;
INPUT NAME $20.;

PROC SORT DATA=TODAY;
BY NAME;

DATA HISTORY;
MERGE DETENT (IN=INA)
      TODAY  (IN=INB);
BY NAME;
IF A AND B;
```

Use a format to eliminate the need to sort:

```
DATA TODAY;
INFILE RAWDATA;
INPUT NAME $20.;
RETAIN FMTNAME '$TODAY'  LABEL='ON THE LIST';
```

```
PROC FORMAT CNTLIN=TODAY (RENAME=(NAME=START));

DATA HISTORY;
SET DETENT;
IF PUT(NAME, $TODAY.) = 'ON THE LIST';
```

The CNTLIN option in PROC FORMAT uses a SAS data set as input instead of a VALUE statement. PROC FORMAT specifically looks for three key variable names: FMTNAME (the name of the format), START (the incoming value), and LABEL (the formatted value).

In a way, this last approach is unfair. After all, this book never described how to create a format from a SAS data set, nor how to utilize a format with the PUT function. While these techniques are important, there is also a more general lesson to be learned here. The more you know, the more choices you have. The more choices you have, the more opportunity you have to select a faster programming technique. Chapter 9 expands on this theme.

CHAPTER 5

Sorting Data

Overview

Because sorting is expensive, it pays to know:

- when the program can process unsorted data

- what alternatives to sorting exist

- how to speed up sorting.

Legend for Icons

CPU time or storage space
Black stars = technique applies frequently
White stars = technique applies infrequently

3 black or white stars - maximum savings
2 black or white stars - medium savings
1 black or white star - smallest savings

Fluctuating savings
Savings can range from large to
nonexistent to negative, depending
on your operating system
or data characteristics.

Archaic
No savings under current
release.

This chapter addresses all three topics. In addition, the final section of this chapter addresses using extra CPU time for sorting, to make up for a shortage of other resources.

Section 1: Do You Need Sorted Data?

For any of these reasons, a program requires sorted data:

- A DATA or PROC step uses a BY statement. (The one exception to this rule, the NOTSORTED option, appears later in this chapter.) Other statements, such as CLASS and TABLES, do not require sorted data.

- PROC PRINT should display the observations in a particular order (even when PROC PRINT does not use a BY statement).

- A handful of procedures (notably FREQ and TABULATE) support the ORDER = DATA option, which displays output in the same order as the incoming data.

- A program might overcome limitations in memory or sort work space by adding a PROC SORT and a subsequent BY statement.

Let's begin by investigating when to skip PROC SORT.

Skip Unnecessary Sorts

The savings are easy to calculate: 100%. Programs that require sorted data do not necessarily require PROC SORT. If the data are already sorted, there is no need to run PROC SORT. Here is a simple example where the second sort is unnecessary:

```
PROC SORT DATA=SALES;
BY SALESMAN AMOUNT;
```

```
DATA TYPICAL;
SET SALES;
BY SALESMAN;
IF FIRST.SALESMAN OR LAST.SALESMAN THEN DELETE;

PROC SORT DATA=TYPICAL;
BY SALESMAN;
PROC MEANS DATA=TYPICAL;
BY SALESMAN;
VAR AMOUNT;
```

Observations leave the DATA step in the same order they enter. Because the incoming data from SALES are sorted BY SALESMAN, and because the DATA step never changes the value of SALESMAN, the outgoing data in TYPICAL are also sorted BY SALESMAN. Deleting an observation here and there does not change that result. Therefore, the second PROC SORT is unnecessary.

Beginning with Release 6.07 of the SAS System, the software stores information about the sorted order of SAS data sets. In that way, the software automatically skips sorts that it "knows" to be unnecessary. In this program, for example, the second PROC SORT gets skipped, while the third gets replaced by copying instead of sorting:

```
PROC SORT DATA=SALES;
BY STATE POP;

PROC SORT DATA=SALES;
BY STATE;

PROC SORT DATA=SALES OUT=SALES2;
BY STATE;
```

Notes appear in the SAS log documenting the actions taken by the software. For the second PROC SORT, the note says:

```
Input data set is already sorted, no sorting done.
```

For the final PROC SORT, the note says:

```
Input data set is already sorted, it has been copied to the
output data set.
```

In both cases, the software "knew" that the incoming data were already sorted and could therefore eliminate unnecessary sorting. There are two methods by which the software "knows" the data are sorted. The first is to run PROC SORT. The second is to add the data set option SORTEDBY=. Consider how this would look in the earlier example, where the second PROC SORT could have been skipped:

```
PROC SORT DATA=SALES;
BY SALESMAN AMOUNT;

DATA TYPICAL (SORTEDBY=SALESMAN AMOUNT);
SET SALES;
BY SALESMAN;
IF FIRST.SALESMAN OR LAST.SALESMAN THEN DELETE;

PROC SORT DATA=TYPICAL;
BY SALESMAN;

PROC MEANS DATA=TYPICAL;
BY SALESMAN;
VAR AMOUNT;
```

Now the software skips the second PROC SORT because it "knows" that TYPICAL is already sorted. BE CERTAIN WHEN USING SORTEDBY=. If you are wrong (meaning the data do not leave the DATA step in order), the DATA step still executes without any error message. The result: data are not sorted, but the software "knows" they are sorted and thus will skip later attempts to run PROC SORT.

Even in obvious cases, you must add the SORTEDBY= data set option to tell the SAS System that the data are sorted. This program, for example, still runs the second PROC SORT:

```
PROC SORT DATA=USA;
BY STATE;

DATA INORDER;
SET USA;

PROC SORT DATA=INORDER;
BY STATE;
```

The software still runs the second PROC SORT because there is no guarantee that INORDER is sorted! For example, this DATA step is perfectly legal:

```
DATA INORDER;
SET USA;
BY STATE;
IF AMOUNT > 20 THEN STATE=' ';
```

The data enter the DATA step in order, but leave out of order. Because this possibility exists, the software never assumes a sorted order for the outgoing data. You must specify the SORTEDBY= data set option if you know the data are sorted. The DATA step would become:

```
DATA INORDER (SORTEDBY=STATE);
SET USA;
```

Now the software skips will skip later attempts to sort INORDER by STATE.

In many types of situations, you know the data are sorted, but the SAS System does not. The list includes:

- The DATA step contains a BY statement, as in some of the examples above.

- The DATA step uses DO loops to create the data, instead of reading an existing file.

```
DATA DOLOOPS (SORTEDBY=I J);
DO I=1 TO 10;
   DO J=1 TO 5;
      OUTPUT;
   END;
END;
```

- The DATA step inputs data from a sorted raw data file.

```
DATA SALES (SORTEDBY=STATE);
INFILE RAW /* Known to be sorted */;
INPUT STATE $2. POP COMMA9.;
```

- The DATA step deletes observations from a sorted data set.

```
DATA VERMONT (SORTEDBY=ZIP);
SET USA;
BY ZIP;
IF STATE='VT';
```

- The DATA step creates multiple data sets or multiple observations from sorted data.

```
DATA NOBOYS   (SORTEDBY=SCHOOL)
     NOGIRLS  (SORTEDBY=SCHOOL)
     BOTH     (SORTEDBY=SCHOOL);
MERGE MEN    (IN=M)
      WOMEN  (IN=W);
BY SCHOOL;
IF      M=0 THEN OUTPUT NOBOYS;
ELSE IF W=0 THEN OUTPUT NOGIRLS;
ELSE                OUTPUT BOTH;

DATA MANYOBS (SORTEDBY=SALESMAN);
SET SALES;
BY SALESMAN;
ARRAY ALLYEAR {4} SALESQ1-SALESQ4;
IF SALARY > 60000 THEN DO QUARTER=1 TO 4;
   SALES = ALLYEAR{QUARTER};
   OUTPUT;
END;
```

As long as you are certain that the new data set is sorted, always code the SORTEDBY= data set option!

Besides skipping unnecessary sorts, two additional benefits can come into play when the SAS System "knows" that a data set is sorted:

- PROC SQL can skip unnecessary internal sorts when joining data sets. Test programs for Chapter 4 revealed that PROC SQL can join sorted data sets much faster. In fact, for test cases, the combination of sorting and then joining generally completed faster than joining unsorted data sets.

- When a WHERE statement is used in combination with a BY statement and an index exists that could retrieve the observations in order, the SAS System would normally use the index to retrieve the observations. However, when the data are known to be sorted, the SAS System retrieves the observations sequentially instead, ignoring the index.

Sort Once, Using the Most Detailed Order

This program sorts one time too many:

```
PROC SORT DATA=SALES;
BY COMPANY;

PROC PRINT DATA=SALES;
BY COMPANY;

PROC SORT DATA=SALES;
BY COMPANY YEAR;

PROC MEANS DATA=SALES;
BY COMPANY YEAR;
VAR AMOUNT;
```

Because the BY statements are very similar, sort in the most detailed order first to eliminate the second PROC SORT:

```
PROC SORT DATA=SALES;
BY COMPANY YEAR;

PROC PRINT DATA=SALES;
BY COMPANY;

PROC MEANS DATA=SALES;
BY COMPANY YEAR;
VAR AMOUNT;
```

If you plan on storing permanent SAS data sets, consider what order would be most useful for later analysis and store data permanently in that order. If you need the data in some other order, add OUT= in later PROC SORTs to preserve the order in the permanent data set:

```
PROC SORT DATA=PERM.SALES OUT=TEMP;
BY UNUSUAL ORDER;
```

The NOTSORTED Option

NOTSORTED processes data that are grouped but not sorted in either ascending or descending order. For the following program to run properly, the data might be grouped, with all the Vermont observations first, all the Maine observations next, and all the Washington observations last.

```
PROC PRINT DATA=SALES;
BY STATE NOTSORTED;
VAR AMOUNT;
```

PROC PRINT assumes that each time STATE changes, the current state is complete. The procedure merely prints a new heading line and moves on to the next STATE without requiring data in ascending or descending order. So even though the data are not sorted, there is not need to run PROC SORT. Of course, you get a meaningless report if you are wrong and the data are not actually grouped. The program merely executes, printing a new block of observations each time STATE changes, without generating an error message.

The potential savings are large (100% of the PROC SORT). You can eliminate PROC SORT entirely whenever your data truly are grouped but not sorted. On the other hand, this situation just doesn't occur very often. The most likely candidate is SAS data being transferred between an IBM mainframe (using the EBCDIC collating sequence) and another platform (using the ASCII collating sequence). In that case, data sorted on one platform are unlikely to be sorted on the other, but they are sure to be grouped. Even under this scenario, the data could have been sorted differently on the original platform. Setting the SORTSEQ= option to EBCDIC or ASCII will utilize the collating sequence of your choice:

```
OPTIONS SORTSEQ=ASCII;

PROC SORT DATA=COMPUTER.USAGE;
BY USERID;
```

The SORTSEQ option introduces one final reason to add NOTSORTED. The following PROC MEANS would generate an error message due to the unusual sorted order:

```
OPTIONS SORTSEQ=DANISH;

PROC SORT DATA=COMPUTER.USAGE;
BY USERID;

PROC MEANS DATA=COMPUTER.USAGE;
VAR CPU_TIME;
BY USERID;
```

Although PROC SORT respects the SORTSEQ option, other procedures do not. PROC MEANS still expects to find the observations in ASCII or EBCDIC order, depending on the operating system. To avoid that error condition, the final statement must read:

```
BY USERID NOTSORTED;
```

Section 2: Alternatives to Sorting

When PROC MEANS (or PROC SUMMARY) summarizes subgroups of the data, it runs fastest against sorted data. However, when the data are not previously sorted, the program as a whole runs faster by skipping PROC SORT. Here is the alternative method.

Know When to Summarize with CLASS Instead of BY

When you summarize data with either PROC MEANS or PROC
SUMMARY, these programs print exactly the same information.
Program #1:

```
PROC MEANS DATA=SALES;
CLASS COMPANY YEAR;
VAR AMOUNT;
```

Program #2:

```
PROC SORT DATA=SALES;
BY COMPANY YEAR;

PROC MEANS DATA=SALES;
BY COMPANY YEAR;
VAR AMOUNT;
```

The first program runs faster but requires more memory. Here are the
details.

If the data were already sorted, the BY statement would run faster
than CLASS. But because CLASS eliminates the need to sort, it makes
the first program faster. On the other hand, the memory requirements
increase dramatically. The first program must hold statistics in memory
for each COMPANY YEAR combination. PROC MEANS in the second
program processes just one COMPANY YEAR combination at a time.
With 1000 companies and 10 years, then, the first PROC MEANS
requires 10,000 times the memory that the second one requires to
hold calculated statistics. In addition, the CLASS statement faces a
theoretical limit of 32,767 different combinations that the CLASS
variables can take on. The BY statement can process unlimited
numbers of combinations.

For the speediest results, follow this general rule: Use BY if the data are already sorted. Otherwise use CLASS. Break that rule for two reasons:

- If memory limitations preclude using CLASS.

- If you need additional levels of summarization stored in an output data set.

Although these programs print the same numbers, if an OUTPUT statement were added, the output data sets would be different. Consider this statement within a PROC MEANS:

```
OUTPUT OUT=STATS SUM=TOTAMOUNT;
```

In combination with the BY statement, this OUTPUT statement would generate one observation for each COMPANY YEAR combination. In combination with the CLASS statement, on the other hand, this OUTPUT statement generates four levels of summarization:

- Each COMPANY YEAR combination (same as for the BY statement).

- Each COMPANY (summing across YEAR values).

- Each YEAR (summing across COMPANY values).

- The entire data set as a whole (summing across all observations).

Chapter 6 explains more of the details and strategies related to multiple levels of summarization.

Section 3: Speeding Up PROC SORT

If you cannot avoid sorting, and if you cannot replace sorting with faster alternatives, you may be able to speed up PROC SORT. Here are some possibilities.

The SORTSIZE= Option

The SORTSIZE= option controls the amount of memory available for sorting. This option can dramatically affect CPU time. While you can specify SORTSIZE= in bytes (as above) or in kilobytes such as SORTSIZE=8000K, the easy way is:

```
PROC SORT DATA=TEST SORTSIZE=MAX;
BY SORTKEY;
```

The keyword MAX uses all available memory.

The reduction in CPU time will vary depending on the amount of memory available and the amount currently being used by PROC SORT. As a result, you will have to run your own tests. But be prepared for savings of 0 to 50%!

The NOEQUALS Option

Without the NOEQUALS option, the order of the observations remains unchanged within a BY group:

```
PROC SORT DATA=SALES;
BY STATE;
```

In this example, the order of the observations within a STATE is the same before and after sorting. The SORT procedure expends CPU time to guarantee that the order will remain the same within each STATE. If you don't care about the order within each STATE, use the NOEQUALS option:

```
PROC SORT DATA=SALES NOEQUALS;
BY STATE;
```

This option eliminates the requirement that the order remain the same within each STATE, enabling PROC SORT to run as much as 15% faster. Some sorting routines ignore the NOEQUALS option.

KEEP= and DROP= Data Set Options

PROC SORT can add the KEEP= (or DROP=) data set option. However, in Version 6 of the software, moving these options from the OUT= data set to the DATA= data set does not help. Consider, for example, these variations on PROC SORT:

```
PROC SORT DATA=HUGE.DATABASE (KEEP=NAME RANK SERIAL)
          OUT=SMALL;
BY NAME;

PROC SORT DATA=HUGE.DATABASE
          OUT=SMALL (KEEP=NAME RANK SERIAL);
BY NAME;
```

In Version 6, both PROC SORTs process all the variables. Regardless of where the KEEP= data set option appears, PROC SORT limits the variables in the output data set only.

Version 7 lets you limit incoming variables by applying the KEEP= data set option to the incoming data set. The global option SORTDUP lets you specify the Version 6 approach (limit variables on the output only) or the faster approach (limit variables from the incoming data set when the option appears on that data set). The reason the software supports both approaches is because the NODUPS option deletes different observations, depending on whether the software locates duplicates based on the subset of variables being read in or based on all variables in the original data set.

Under any release of the software, be wary of this final variation. It replaces the original permanent data set with a much smaller one, containing only three variables:

```
PROC SORT DATA=HUGE.DATABASE (KEEP=NAME RANK SERIAL);
BY NAME;
```

Sort in Ascending Order

Sorting in ascending order runs marginally faster than sorting in descending order. Some applications give you no choice; the programming logic requires one particular order. But for applications that could use either order, use ascending order.

Selecting the Sorting Routine

The SAS software contains a sorting routine that works quickly on relatively small data sets. If your operating system also provides a sorting utility, systems programmers can make both sorting tools available when installing the software. You won't have to choose the faster routine: the software will choose for you on each PROC SORT.

Typically, the SAS sorting routine works faster on smaller data sets. Thus, the software automatically uses the SAS sorting routine for small data sets and (if available) the system sorting routine for larger data sets. Two options let you modify this process. First, the SORTCUTP option lets you control what the software considers "small" and what it considers "large." Second, the SORTPGM option lets you request a particular sorting routine. Valid values include:

```
OPTIONS SORTPGM=SAS;  /* use the SAS sorting tool    */
OPTIONS SORTPGM=HOST; /* use the host sorting utility */
OPTIONS SORTPGM=BEST; /* calculate which runs faster  */
```

Surprisingly, tests on OpenVMS found the SAS sorting routine running 35% faster than the host sort utility. The test data set contained 100,000 observations and 100 variables.

One more possibility is that you may be able to sort your raw data with another software tool. That approach is often faster than running PROC SORT. Also note that some sorting software can take subsets of the data, calculate sums, or produce frequency counts from raw data.

Section 4: Trading Off Resources

Finally, let's consider a couple of cases where speed is not of the essence. In these situations, other resources are in short supply. By expending additional CPU time for PROC SORT, the program manages to overcome a shortfall in some other resource.

Trading CPU Time for Memory

Earlier, this chapter notes how the CLASS statement forces PROC MEANS to hold statistics in memory for each combination of the CLASS variables. By switching to a BY statement (requiring sorted data), the program could use extra CPU time to overcome memory limitations. However, memory limitations can apply to other procedures as well. Both PROC FREQ and PROC TABULATE, for example, store statistics in memory for each cell in the table. This program, for example, could run out of memory because the page dimension multiplies by 50 the number of cells in the table:

```
PROC FREQ DATA=SALES;
TABLES STATE*SALESMAN*YEAR;
```

PROC FREQ must hold counts for each STATE/SALESMAN/YEAR combination in memory at the same time. Even if the operating system imposes no memory limitations, the SAS System will not track more than 32,767 levels of the TABLES variables at once. If memory is the limiting resource, then, your program can trade CPU time for memory usage:

```
PROC SORT DATA=SALES;
BY STATE;

PROC FREQ DATA=SALES;
TABLES SALESMAN*YEAR;
BY STATE;
```

The two PROC FREQs produce identical tables, except for two slight differences:

- If a STATE contains no sales in one YEAR, the first set of tables will contain a column with all zeros (but the second set will not).

- If table options request overall statistics, those will be different.

These differences are inconsequential for most PROC FREQs. By adding the BY statement to process each STATE separately, the second PROC FREQ needs just 2% of the memory needed by the first to hold frequency counts.

Trading CPU Time for Sort Workspace

When sorting a large data set, the necessary sort workspace may be unavailable. At the expense of using more CPU time, the program can decrease the need for sort workspace. These are measures of last resort, when your back is to the wall because PROC SORT cannot complete. These measures can dramatically increase CPU time.

One method is the TAGSORT option:

```
PROC SORT DATA=SALES TAGSORT;
BY STATE COUNTY;
```

This option doesn't read in the entire observation into the sort work areas. It reads in the BY variables and observation number and sorts only that information. Then it uses the observation number to retrieve the remaining variables from the original data.

TAGSORT is very slow! One reason for this slowdown is that TAGSORT automatically invokes the SAS sorting routine (as opposed to SYNC-SORT or other sorting routines available on your computer system), which is inefficient for large numbers of observations. Thus, the major usefulness of this option occurs for data sets with a large number of variables, rather than a large number of observations.

Writing your own version of TAGSORT can work faster than the PROC SORT option. The program below follows the same strategy of sorting the observation number and then retrieving the remaining variables:

```
DATA JUSTKEYS;
SET ORIGINAL (KEEP=STATE COUNTY);
OBSNO=_N_;

PROC SORT DATA=JUSTKEYS;
BY STATE COUNTY;

DATA ORIGINAL (DROP=OBSNO);
SET JUSTKEYS (KEEP=OBSNO);
SET ORIGINAL POINT=OBSNO;
```

The POINT= option in the SET statement examines the value of OBSNO and retrieves that particular observation from ORIGINAL. Chapter 10 reports the test results for this workaround program, which ran much faster under the MVS operating environment but ran marginally slower on other operating systems.

On a PC, the SAS System uses available space on the hard drive to sort the data and to hold the final sorted data set. On a mainframe, however, the sort work areas and the work area occupy separate blocks of storage space. Also, the space needed to sort a data set is larger than the space needed to hold the final sorted data. Thus, when sort workspace is in short supply, a program might break up a data set into smaller subsets, sort the subsets, and then reassemble the subsets into one large data set. Consider the case where SALES is too large to be sorted:

```
PROC SORT DATA=SALES;
BY STATE COUNTY;
```

First, break up SALES into separate data sets:

```
DATA SUBSET1 SUBSET2 SUBSET3;
SET SALES NOBS=TOTOBS;
IF      _N_ <    TOTOBS/3 THEN OUTPUT SUBSET1;
ELSE IF _N_ < 2*TOTOBS/3 THEN OUTPUT SUBSET2;
ELSE                          OUTPUT SUBSET3;
```

Next, run a separate PROC SORT on each subset.

```
PROC SORT DATA=SUBSET1;
BY STATE COUNTY;

PROC SORT DATA=SUBSET2;
BY STATE COUNTY;

PROC SORT DATA=SUBSET3;
BY STATE COUNTY;
```

The program uses less sort workspace because it sorts a smaller data set at any one point in time. The program can reuse the same sort work space to sort each of the data sets. Finally, reassemble all data sets into one:

```
DATA SALES;
SET SUBSET1 SUBSET2 SUBSET3;
BY STATE COUNTY;
```

The total CPU time rises. The total amount of workspace rises temporarily, until the program subsequently deletes SUBSET1, SUBSET2, and SUBSET3. But the sort workspace requirements decrease because the program sorts relatively small data sets.

The program can minimize the extra CPU time when the primary sort keys take on a known set of values. Consider how the program could proceed if SALES contained just three values for STATE:

```
DATA SALES MA NC;
SET SALES;
IF      STATE='AL' THEN OUTPUT SALES;
ELSE IF STATE='MA' THEN OUTPUT MA;
ELSE /* STATE='NC' */   OUTPUT NC;
```

Because each data set holds data for a different STATE, the program can sort by one variable instead of two:

```
PROC SORT DATA=SALES;
BY COUNTY;
```

```
PROC SORT DATA=MA;
BY COUNTY;

PROC SORT DATA=NC;
BY COUNTY;

PROC APPEND BASE=SALES DATA=MA;
PROC APPEND BASE=SALES DATA=NC;
```

With all these extra steps, the new program will not run faster than a single PROC SORT. However, when a single PROC SORT becomes impossible because of a lack of sort workspace, longer running programs become a viable alternative.

A Small Workshop

Reduce the resources used for sorting in the program below. You can change the order of the reports if that would be helpful.

```
PROC SORT DATA=SALES;
BY STATE;

PROC MEANS DATA=SALES;
BY STATE;
VAR AMOUNT;

PROC SORT DATA=SALES;
BY SALESMAN;

PROC FREQ DATA=SALES;
TABLES SALESMAN;

PROC SORT DATA=SALES;
BY STATE YEAR;

PROC UNIVARIATE DATA=SALES;
BY STATE YEAR;
VAR AMOUNT;
```

Workshop Solution

You can eliminate two out of the three PROC SORTs:

PROC FREQ works just as quickly on unsorted data as on sorted data. The second PROC SORT is unnecessary.

The first and last sorts are very similar. Run the more detailed PROC SORT first, to eliminate the need for two sorts.

The program becomes:

```
PROC SORT DATA=SALES;
BY STATE YEAR;

PROC MEANS DATA=SALES;
BY STATE;
VAR AMOUNT;

PROC FREQ DATA=SALES;
TABLES SALESMAN;

PROC UNIVARIATE DATA=SALES;
BY STATE YEAR;
VAR AMOUNT;
```

The order of the reports can actually remain the same. Changing PROC MEANS to use CLASS instead of BY would be a mistake:

```
PROC MEANS DATA=SALES NWAY;
CLASS STATE;
VAR AMOUNT;
```

The CLASS statement is slower than the BY statement. Because PROC UNIVARIATE does not support a CLASS statement, the program must run PROC SORT. In that case, use the BY statement in PROC MEANS to take advantage of the fact that the data are sorted.

CHAPTER 6

Summarizing Data

Overview

When summarizing data, efficiency takes two forms:

- Selecting the fastest method of generating a particular summary.

- Planning to permanently save summary data sets, so that later programs can reuse the summaries instead of analyzing the original data.

Legend for Icons

CPU time or storage space
Black stars = technique applies frequently
White stars = technique applies infrequently
3 black or white stars - maximum savings
2 black or white stars - medium savings
1 black or white star - smallest savings

Fluctuating savings
Savings can range from large to nonexistent to negative, depending on your operating system or data characteristics.

Archaic
No savings under current release.

This chapter contains one section on each of these topics. The second topic represents a wave of the future in the data processing world. In many ways, the SAS System is moving in the direction of reusing summary data sets:

- In Version 7, most procedures can create output data sets.

- Executive information systems extract, update, and summarize data overnight, so the data will be available on the desktop the next morning.

- Data warehousing involves regular, massive summarizations, with the results widely available for further inspection and analysis.

This chapter narrows down the broad spectrum of summary data sets, focusing on efficient use of PROC SUMMARY and PROC FREQ. For comparison purposes, this chapter examines DATA steps and PROC SQL, which can also calculate summary statistics. In that light, this chapter presents only a sampling of the tools and techniques that summarize data.

Section 1: Faster Summaries

Chapter 4 examined summarizing with a CLASS statement to avoid PROC SORT. Here, we'll consider a few more techniques to generate summaries faster.

Add a Format, Save a DATA Step

Formats can categorize a set of values, eliminating an entire DATA step from the program. Consider this program that fails to use a format:

```
DATA NEW;
SET OLD;
```

```
LENGTH INITIAL $ 1 CATEGORY $ 6;
INITIAL = LASTNAME;
IF      MRATE <  7 THEN CATEGORY='LOW';
ELSE IF MRATE < 10 THEN CATEGORY='MEDIUM';
ELSE                    CATEGORY='HIGH';

PROC FREQ DATA=NEW;
TABLES INITIAL CATEGORY;
```

The proper formats can eliminate the DATA step entirely:

```
PROC FORMAT;
VALUE RATE  LOW - < 7 = 'LOW'
              7 - <10 = 'MEDIUM'
             10 - HIGH = 'HIGH';

PROC FREQ DATA=OLD;
TABLES LASTNAME MRATE;
FORMAT LASTNAME $ 1.  MRATE RATE.;
```

PROC FREQ automatically prints one row for each formatted value of the incoming variable, not for each actual value. Similar considerations apply to PROC SUMMARY. Without a format, the program might use an identical DATA step to set up proper values for CATEGORY, followed by:

```
PROC SUMMARY DATA=NEW PRINT NWAY;
CLASS CATEGORY;
VAR AMOUNT;
```

With a format, however, PROC SUMMARY can process the original data set and variables:

```
PROC SUMMARY DATA=OLD PRINT NWAY;
CLASS MRATE;
VAR AMOUNT;
FORMAT MRATE RATE.;
```

Be careful if you apply a format to a BY variable in PROC SUMMARY. In this program, the CLASS statement combines each formatted value of CATEGORY into one line of the report:

```
PROC FORMAT;
VALUE ODDEVEN 1,3='ODD'  2,4='EVEN';

PROC SUMMARY DATA=SALES PRINT NWAY;
CLASS CATEGORY;
* vs. BY CATEGORY;
FORMAT CATEGORY ODDEVEN.;
VAR AMOUNT;
```

Assuming that CATEGORY always ranges from 1 through 4, the maximum number of lines is one for CATEGORY=ODD and one for CATEGORY=EVEN. If you switch to a BY statement instead, you can get as many as four lines in the report! The BY statement groups CONSECUTIVE formatted values (or ranges) only. So if CATEGORY actually took on values of 1, 2, and 3, you would get three lines in the report; if it took on values of 1, 2, and 4, you would get just two lines because the consecutive values of 2 and 4 would be grouped together as one.

Replacing a Procedure with a DATA Step

SAS procedures are wonderful. They work, they are debugged, and they require very little coding. On the other hand, if you are willing to take the time to code in a DATA step, you may be able to replace a longer-running procedure.

In this example, the SAS data set VISITS contains one observation per visit. The program must count the number of patients with 1 to 3 visits, 4 to 6 visits, 7 to 9 visits, and 10 or more visits.

```
PROC SORT DATA=VISITS;
BY PATIENT;

DATA _NULL_;
SET VISITS (KEEP=PATIENT) END=NOMORE;
BY PATIENT;
COUNT + 1;
```

```
IF LAST.PATIENT;
IF        COUNT <= 3 THEN COUNT1 + 1;   /* 1-3 visits */
ELSE IF COUNT <= 6 THEN COUNT2 + 1;   /* 4-6 visits */
ELSE IF COUNT <= 9 THEN COUNT3 + 1;   /* 7-9 visits */
ELSE                      COUNT4 + 1;   /* 10+ visits */
COUNT = 0;
IF NOMORE THEN DO;
    FILE PRINT;
    PUT /// COUNT1 3.  ' Patients with 1-3 visits'
          // COUNT2 3.  ' Patients with 4-6 visits'
          // COUNT3 3.  ' Patients with 7-9 visits'
          // COUNT4 3.  ' Patients with 10+ visits';
END;
```

Although the program is somewhat lengthy, a single DATA step can use sorted data to count visits, categorize the number of visits, and print the final report. In contrast, Section 2 of this chapter approaches this problem in a different way (resummarizing an output data set from PROC FREQ). In terms of CPU time, the key factor may be whether the data are already sorted, eliminating the need for PROC SORT. Remember, though, that the extra programming time to write a lengthy DATA step represents another resource that has its own set of costs.

Basic Statistics: Comparisons of Alternative Methods

PROC MEANS, DATA steps, and PROC SQL can calculate essentially the same statistics. (PROC SQL does not calculate weighted statistics or a few of the less common statistics such as skewness and kurtosis.) We will compare just a handful of situations: calculating SUM and N for an entire data set and for subgroups.

These three alternatives print SUM and N for an entire data set:

```
PROC MEANS DATA=SALES SUM N;
VAR AMOUNT;

PROC SQL;
SELECT SUM(AMOUNT) AS SUM, N(AMOUNT) AS N FROM SALES;
```

```
DATA _NULL_;
SET SALES (KEEP=AMOUNT) END=EOF;
SUM + AMOUNT;
IF AMOUNT > .Z THEN N + 1;
IF EOF;
FILE PRINT;
PUT SUM= N=;
```

In test cases, the DATA step ran 20% to 50% faster than PROC SQL. PROC MEANS ranged from being the fastest to the slowest method, depending on the operating environment. See Chapter 10 for details. When considering whether to use the DATA step, weigh these factors as well:

- You have to write somewhat more complex code for a DATA step approach, but procedures invoke written, tested code. In these examples, it would have been easy to overlook special missing values in calculating the N statistic. The procedures would never do that.

- The program may require a DATA step for other reasons. In that case, adding the calculation of SUM and N to the DATA step would use extremely little additional CPU time.

Finally, let's calculate those same statistics for subgroups. These are the comparison programs:

```
PROC MEANS DATA=SALES;
CLASS STATE;
VAR AMOUNT;

PROC SQL;
SELECT SUM(AMOUNT) AS SUM, N(AMOUNT) AS N FROM SALES
GROUP BY STATE;

PROC MEANS DATA=SALES;
BY STATE;
VAR AMOUNT;
```

```
DATA _NULL_;
SET SALES (KEEP=STATE AMOUNT);
BY STATE;
SUM + AMOUNT;
IF AMOUNT > .Z THEN N + 1;
IF LAST.STATE;
FILE PRINT;
PUT STATE= MEAN= N=;
SUM=0;
N=0;
```

The relative speed varied sharply depending on the operating system. Again, Chapter 10 presents the details. The first two methods were tested on unsorted data. PROC MEANS with a CLASS statement ran at least 30% faster than PROC SQL on most operating environments. However, PROC MEANS took twice as long under OpenVMS. Such substantial differences indicate you must perform your own tests.

Three methods were tested on sorted data. In every case, the CPU time for sorting exceeded that for PROC MEANS with a CLASS statement (running against unsorted data). So rule number one is to avoid sorting just to be able to summarize subgroups. However, for data that have already been sorted, consider three methods: PROC MEANS with a BY statement, PROC SQL (retested here because PROC SQL runs faster once the data have been sorted), and a DATA step. Again, Chapter 10 shows how the results vary dramatically with the operating system. While the DATA step usually ran fastest, under Windows NT PROC MEANS completed in half the time. Once again, it pays to perform your own tests.

PROC SUMMARY vs. PROC MEANS

PROC SUMMARY and PROC MEANS have exactly the same capabilities and run with exactly the same efficiency. Historically, this has not always been the case; it is true beginning with Version 6 of the software, however. The procedures have slightly different default actions. For example, PROC MEANS automatically prints its results (although that can be suppressed with the NOPRINT option). PROC SUMMARY does not print its results (unless you add the PRINT option).

Section 2: Reusing Summary Data Sets

Obviously, processing a small summary data set takes less CPU time than processing the original data. To make this happen, you need two key skills. First, you must understand the structure of summary data sets. Second, you must develop your file handling skills so that your programs can flexibly and easily read summaries. This section illustrates these principles, using PROC SUMMARY and PROC FREQ.

The Structure of PROC SUMMARY Output Data Sets

PROC SUMMARY output data sets automatically include all variables you create with the OUTPUT statement, the automatic variables _TYPE_ and _FREQ_, each BY variable, and each CLASS variable. If PROC SUMMARY uses an ID statement, the ID variables will also appear in the output data set. Here is a typical example:

```
PROC SUMMARY DATA=ALL.SALES NWAY;
CLASS DEPT JOBGRADE;
VAR SALARY;
OUTPUT OUT=SMALL.STATS SUM=TOTSAL N=EMPLOYES;
```

The output data set contains TOTSAL, EMPLOYES, _TYPE_, _FREQ_, DEPT, and JOBGRADE. There is one observation for each DEPT/JOBGRADE combination except for missing values, which would (without the MISSING option) be discarded from the output data set.

The program could generate additional levels of summarization by removing the NWAY option from the PROC statement:

```
PROC SUMMARY DATA=ALL.SALES;
CLASS DEPT JOBGRADE;
VAR SALARY;
OUTPUT OUT=SMALL.STATS SUM=TOTSAL N=EMPLOYES;

PROC PRINT DATA=SMALL.STATS;
FORMAT TOTSAL COMMA7.;
```

The results:

OBS	DEPT	JOBGRADE	_TYPE_	_FREQ_	EMPLOYES	TOTSAL
1		.	0	15	15	485,000
2		8	1	3	3	85,000
3		9	1	8	8	265,500
4		10	1	4	4	134,500
5	ENGINEERING	.	2	8	8	283,100
6	SALES	.	2	7	7	201,900
7	ENGINEERING	8	3	1	1	24,000
8	ENGINEERING	9	3	4	4	163,100
9	ENGINEERING	10	3	3	3	96,000
10	SALES	8	3	2	2	61,000
11	SALES	9	3	4	4	102,400
12	SALES	10	3	1	1	38,500

With the NWAY option, the output data set would have contained the last six observations only, representing each DEPT/JOBGRADE combination. Removing the NWAY option adds observations 1 through 6 to the output data set, representing other levels of summarization. The values for _TYPE_ indicate the level of summarization by answering a key question for each CLASS variable: Does this observation represent a separate analysis for each DEPT? For each JOBGRADE? The following table lists all possible answers to these questions, along with the corresponding value of _TYPE_.

DEPT?	JOBGRADE?	_TYPE_
NO	NO	00 = 0
NO	YES	01 = 1
YES	NO	10 = 2
YES	YES	11 = 3

Each set of YES and NO answers matches a level of summarization. For example, answering YES for DEPT and NO for JOBGRADE requests a separate set of statistics for each DEPT but not for each JOBGRADE (lump all JOBGRADEs together within the DEPT). This corresponds to observations 5 and 6 in the output data set, the only observations where _TYPE_ has a value of 2. Answering NO for JOBGRADE and NO for DEPT requests statistics for all JOBGRADEs and DEPTs lumped together. This corresponds to observation 1 in the output data set, the only observation where _TYPE_ has a value of 0.

Somehow, _TYPE_ is identifying the level of summarization. PROC SUMMARY assigns values to _TYPE_ by converting each YES to a 1 and each NO to a 0, then interpreting the combination of 1s and 0s as a binary integer.

The other automatic variable, _FREQ_, is the total number of observations in the DEPT/JOBGRADE category. That differs from EMPLOYES, which is the number of observations having a nonmissing SALARY. Similarly, the NMISS statistic would be the number of observations having a missing value for SALARY. So _FREQ_ is actually NMISS + N.

When the program omits NWAY, the output data set will contain many different levels of summarization and correspondingly many values for _TYPE_. With a CLASS statement like the following, how can you select the proper level of summarization from the output data set?

```
CLASS COUNTRY DIVISION DEPT JOBGRADE;
```

Using a binary literal representation makes the process a little easier. For example, these two statements ask for the same observations:

```
IF _TYPE_=5;
IF _TYPE_='0101'B;
```

Match up the 0s and 1s to the variables in the CLASS statement. The variables that match to a 1 change, while those which match to a 0 do not change. So this subset represents a separate analysis for each DIVISION/JOBGRADE combination, lumping together all values for COUNTRY and DEPT. Note that you cannot use a binary literal in a WHERE statement.

This description of both PROC SUMMARY and its output data set covers the basics only. For a more thorough description, refer to the SAS Institute publication *SAS Procedures Guide, Version 6, Third Edition*, pages 1-26, 365-388, and 559-560. If you are working with a different edition, the documentation appears under three topics: PROC MEANS, PROC SUMMARY, and a separate chapter comparing elementary statistics procedures.

PROC SUMMARY Output Data Sets: Integrating Levels of Summarization

When an output data set contains many levels of summarization, it opens up the door for clever programming techniques. For example, notice that the sorted order to the 12 observations printed above is BY _TYPE_ DEPT JOBGRADE. Therefore, we could obtain a detailed set of observations for each DEPT, followed by a summary for that DEPT, by reading a block of _TYPE_=3 observations followed by the _TYPE_=2 observation for the matching DEPT. This DATA step would do the trick:

```
DATA BYDEPT;
SET SMALL.STATS (WHERE=(_TYPE_=3))
    SMALL.STATS (WHERE=(_TYPE_=2));
BY DEPT;
AVERAGE = TOTSAL / EMPLOYES;
```

Let's expand this concept to work with two summary data sets at once. Begin with two unsummarized data sets (SALES and COSTS), each containing three key variables: STATE, YEAR, and AMOUNT. Each data set contains many observations for each STATE/YEAR combination. The programming objective is to summarize both data sets and combine the observations from these summaries in the following way:

- The order to the final data set should be by STATE: all the detailed (STATE/YEAR) observations for that STATE, followed by the summary observation for that STATE.

- Each detailed observation in the summary should contain STATE, YEAR, the average sales for that STATE/YEAR, and the average cost for that STATE/YEAR.

- Each summary observation should contain STATE, the average sales for that STATE across all years, and the average cost for that STATE across all years.

This diagram outlines the intended result:

```
STATE          YEAR          AVGSALES                AVGCOSTS

STATE #1       YEAR #1       STATE/YEAR average      STATE/YEAR average
STATE #1       YEAR #2       STATE/YEAR average      STATE/YEAR average
STATE #1       YEAR #3       STATE/YEAR average      STATE/YEAR average
STATE #1       (All)         STATE average           STATE average

STATE #2       YEAR #1       STATE/YEAR average      STATE/YEAR average
STATE #2       YEAR #2       STATE/YEAR average      STATE/YEAR average
STATE #2       YEAR #3       STATE/YEAR average      STATE/YEAR average
STATE #2       (All)         STATE average           STATE average

. . .

STATE #n       YEAR #1       STATE/YEAR average      STATE/YEAR average
STATE #n       YEAR #2       STATE/YEAR average      STATE/YEAR average
STATE #n       YEAR #3       STATE/YEAR average      STATE/YEAR average
STATE #n       (All)         STATE average           STATE average
```

The program begins by summarizing both original data sets:

```
PROC SUMMARY DATA=SALES;
CLASS STATE YEAR;
VAR AMOUNT;
OUTPUT OUT=SALESSUM MEAN=AVGSALES;

PROC SUMMARY DATA=COSTS;
CLASS STATE YEAR;
VAR AMOUNT;
OUTPUT OUT=COSTSSUM MEAN=AVGCOSTS;
```

Each summary data set contains four levels of summarization: a summary for the entire data set (_TYPE_=0), a summary for each YEAR (_TYPE_=1), a summary for each STATE (_TYPE_=2), and a summary for each STATE/YEAR (_TYPE_=3). The remainder of the program must reorganize these summaries, subsetting the observations and putting them in the proper order.

Because the program works with summary data sets from this point
forward, it won't use much CPU time by processing the summaries a
few times. Here is one approach that will combine the two summaries:

```
DATA TYPE2;
MERGE SALESSUM COSTSSUM;
BY STATE;
WHERE _TYPE_=2;

DATA TYPE3;
MERGE SALESSUM COSTSSUM;
BY STATE YEAR;
WHERE _TYPE_=3;

DATA COMBINED;
SET TYPE2 TYPE3;
BY STATE;
```

And here is a shorter approach, requiring better file handling skills:

```
DATA COMBINED;
MERGE SALESSUM (WHERE=(_TYPE_=3))
      COSTSSUM (WHERE=(_TYPE_=3));
BY STATE YEAR;
OUTPUT;
IF LAST.STATE THEN DO;
   MERGE SALESSUM (WHERE=(_TYPE_=2))
         COSTSSUM (WHERE=(_TYPE_=2));
   BY STATE;
   OUTPUT;
END;
```

These features of the summaries combine to make the program work:

- The _TYPE_=3 observations are in order by STATE.

- The _TYPE_=2 observations are in order by STATE, with one
  observation per STATE.

- Both sets of observations contain data for exactly the same
  STATEs.

These programs work well with small summary data sets. However, when a few CLASS variables take on many different values, the size of a summary data set can rival the size of the original! For those cases, we'll consider a different strategy: use NWAY to save just the most detailed level of summarization and then resummarize the summary to obtain other levels of summarization as needed.

PROC SUMMARY Output Data Sets: Resummarizing

When PROC SUMMARY uses the NWAY option, the output data set contains only the most detailed level of summarization:

```
PROC SUMMARY DATA=ALL.SALES NWAY;
CLASS DEPT JOBGRADE;
VAR SALARY;
OUTPUT OUT=SMALL.STATS SUM=TOTSAL N=EMPLOYES;

PROC PRINT DATA=SMALL.STATS;
FORMAT TOTSAL COMMA7.;
```

The results:

OBS	DEPT	JOBGRADE	_TYPE_	_FREQ_	EMPLOYES	TOTSAL
1	ENGINEERING	8	3	1	1	24,000
2	ENGINEERING	9	3	4	4	163,100
3	ENGINEERING	10	3	3	3	96,000
4	SALES	8	3	2	2	61,000
5	SALES	9	3	4	4	102,400
6	SALES	10	3	1	1	38,500

If we wanted a summary at the DEPT level, we would have to resummarize this summary data set. For example:

```
PROC SUMMARY DATA=SMALL.STATS NWAY;
CLASS DEPT;
VAR TOTSAL EMPLOYES;
OUTPUT OUT=STATS SUM=;
```

Presumably, the summary SMALL.STATS is relatively small (compared to the original ALL.SALES), so we can process the summary over and over using relatively little CPU time. To easily resummarize the summary, the original PROC SUMMARY should create basic statistics. For example, even if we eventually want the MEAN statistic for each value of DEPT, the original PROC SUMMARY should calculate the SUM and N statistics. The resummarization program would total the SUM and N statistics at the DEPT level and calculate the MEAN from those totals. Along the same lines, by saving and aggregating N, SUM, and USS (the sum of the squared values) at a detailed level, you can calculate standard deviation at an aggregate level using the formula (for unweighted statistics):

```
STD = (USS - SUM**2/N) / (N - 1)
```

The Structure of PROC FREQ Output Data Sets

The variables in a PROC FREQ output data set include any incoming variables, whether in a TABLES or BY statement, as well as the automatic variables COUNT and PERCENT. With additional options not illustrated here, you can also capture marginal counts and percents in an output data set. For details, refer to the SAS Institute publication SAS Technical Report P-222, *Changes and Enhancements to Base SAS Software, Release 6.07*, pages 219-224.

Consider the following cross-tabulation:

```
PROC FREQ DATA=SALARIES;
TABLES JOBGRADE*DEPT / NOPRINT OUT=TOTALS;
```

The SAS data set TOTALS contains one observation for each JOBGRADE/DEPT combination, with the variables JOBGRADE, DEPT, COUNT, and PERCENT. COUNT is the number of observations for that particular JOBGRADE/DEPT combination, while PERCENT is the percent of the entire data set represented by that COUNT. Unless the TABLES statement adds the MISSING option, all PERCENT calculations exclude cells where JOBGRADE or DEPT has a missing value.

As was the case with PROC SUMMARY, this description of PROC FREQ and its output data set covers the basics only. For a more thorough description, refer to the SAS Institute publication *SAS Procedures Guide, Version 6, Third Edition*, pages 325-363.

When PROC FREQ creates an output data set, you gain a couple of advantages. Instead of letting the procedure control the reporting format, you can control the format in a later step. You can also resummarize that output data set. Let's examine that topic in more detail.

PROC FREQ Output Data Sets: Resummarizing

Here is a simple example of resummarizing:

```
PROC FREQ DATA=ALL.CITIES;
TABLES STATE*URBAN / NOPRINT OUT=DETAILS;
```

Assuming that ALL.CITIES contains one observation per city, DETAILS now contains a few observations for each STATE (one for each value of URBAN within that STATE). To obtain the number of cities in each STATE, you could go back to the original data:

```
PROC FREQ DATA=ALL.CITIES;
TABLES STATE;
```

Alternatively, you could resummarize the output data set DETAILS:

```
PROC MEANS DATA=DETAILS SUM;
BY STATE;
VAR COUNT;
```

Notice that PROC MEANS can use the BY statement because PROC FREQ outputs DETAILS in sorted order.

Let's take one more example, revisiting the earlier portion of this chapter, "Replacing a Procedure with a DATA Step". We are looking to find out how many patients had from 1 to 3 visits, how many had from 4 to 6, how many had 7 to 9, and how many had ten or more visits. The program could begin by having PROC FREQ count the number of visits for each patient:

```
PROC FREQ DATA=VISITS;
TABLES PATIENT / NOPRINT OUT=STATS;
```

The program could then continue by regrouping those counts with a format:

```
PROC FORMAT;
VALUE HOWMANY 1-3='1-3 Visits'
              4-6='4-6 Visits'
              7-9='7-9 Visits'
              10-HIGH='10+ Visits';

PROC FREQ DATA=STATS;
TABLES COUNT;
FORMAT COUNT HOWMANY.;
```

The second PROC FREQ processes the smaller, summary data set that contains one observation per PATIENT.

A Small Workshop

This workshop is all about working with summary data sets, not about selecting the most efficient method to summarize. The workshop asks you to solve the same problems twice, using different initial structures to the data.

Problem #1

What percentage of last names are "Jones", what percentage are "Smith", and what percentage are anything else?

> Situation 1: The incoming data are not summarized. Each observation in the data set PEOPLE represents a single individual and contains the variable LASTNAME.

> Situation 2: The incoming data have already been summarized by PROC FREQ as follows:

```
PROC FREQ DATA=PEOPLE;
TABLES LASTNAME / NOPRINT OUT=STATS;
```

Problem #2

Answer all four questions for each PROC SUMMARY below.

1. Assuming 50 states and 2 genders, how many observations are in the output data set?

2. Print the total income for each state.

3. Print the average income for all females and all males.

4. For each state, print the percent of total taxes paid by females and the percent paid by males.

   Situation 1: The incoming data have been summarized at all relevant levels of summarization:

```
PROC SUMMARY DATA=ALL.TAXPAYER;
CLASS STATE    /* two-letter code */
      GENDER   /* F or M          */;
VAR INCOME TAX;
OUTPUT OUT = TOTAL.TAXES
       SUM = T_INCOME T_TAX
       N   = N_INCOME N_TAX;
```

   Situation 2: The incoming data have been summarized at the most detailed level of summarization only.

```
PROC SUMMARY DATA=ALL.TAXPAYER NWAY;
CLASS STATE    /* two-letter code */
      GENDER   /* F or M          */;
VAR INCOME TAX;
OUTPUT OUT = TOTAL.TAXES
       SUM = T_INCOME T_TAX;
       N   = N_INCOME N_TAX;
```

Workshop Solutions

These sample solutions use the same PROCs we have just been examining. It is also possible to answer some of these questions using a DATA step.

Problem 1, Situation 1. One observation per individual.

```
PROC FORMAT;
VALUE $LNAME    'Smith'='Smith'
                'Jones'='Jones'
                 Other='Other';

PROC FREQ DATA=PEOPLE;
TABLES LASTNAME;
FORMAT LASTNAME $LNAME.;
```

Problem 1, Situation 2. One observation per last name.

```
PROC FORMAT;
VALUE $LNAME    'Smith'='Smith'
                'Jones'='Jones'
                 Other='Other';

PROC MEANS DATA=STATS SUM;
VAR PERCENT;
CLASS LASTNAME;
FORMAT LASTNAME $LNAME.;
```

A DATA step approach is particularly viable here as well. Note that the WHERE statement can subset and still accurately detect end-of-file:

```
DATA _NULL_;
SET STATS END=EOF;
WHERE LASTNAME='Smith' OR LASTNAME='Jones';
FILE PRINT;
PUT // LASTNAME   @20 PERCENT 5.1;
TOTPCT + PERCENT;
IF EOF THEN DO;
   OTHERS = 100 - TOTPCT;
   PUT // 'Others' @20 OTHERS 5.1;
END;
```

Problem 2, Situation 1. Summary data set contains all levels of summarization.

1. The data set contains 153 observations: 100 for the state/ gender combinations, 50 for the states, 2 for the genders, and 1 for the entire data set as a whole.

2.
```
PROC PRINT DATA=TOTAL.TAXES;
WHERE _TYPE_=2;
VAR STATE T_INCOME;
```

3.
```
DATA _NULL_;
SET TOTAL.TAXES;
WHERE _TYPE_=1;
AVERAGE = T_INCOME / N_INCOME;
PUT 'Average income for ' GENDER '= ' AVERAGE;
```

4.
```
DATA TEMP (KEEP=STATE GENDER PERCENT);
MERGE TOTAL.TAXES (WHERE=(_TYPE_=3))
      TOTAL.TAXES (WHERE=(_TYPE_=2)
                   RENAME=(T_TAX=TOTAL_ST));
BY STATE;
PERCENT = T_TAX / TOTAL_ST * 100;

PROC PRINT DATA=TEMP;
```

Problem 2, Situation 2. Summary data set contains only the most detailed level of summarization.

1. The data set contains 100 observations, one for each state/gender combination.

2.
```
PROC MEANS DATA=TOTAL.TAXES SUM;
BY /* or CLASS */ STATE;
VAR T_INCOME;
```

The BY statement works because TOTAL.TAXES must be in sorted order. If the program had used CLASS instead of BY, it would have printed the same numbers whether or not the program added the NWAY option. The CLASS statement prints the most detailed level of summarization only. It adds multiple levels of summarization to any output data sets, but not to the printed report.

```
3. PROC SUMMARY DATA=TOTAL.TAXES NWAY;
   CLASS GENDER;
   VAR T_INCOME N_INCOME;
   OUTPUT OUT=TEMP SUM=;

   DATA _NULL_;
   SET TEMP;
   AVERAGE = T_INCOME / N_INCOME;
   PUT 'Average income for ' GENDER '= ' AVERAGE;

4. PROC SUMMARY DATA=TOTAL.TAXES NWAY;
   CLASS STATE;
   VAR T_TAX;
   OUTPUT OUT=STATS SUM=TOTAL_ST;

   DATA TEMP (KEEP=STATE GENDER PERCENT);
   MERGE TOTAL.TAXES STATS;
   BY STATE;
   PERCENT = T_TAX / TOTAL_ST * 100;

   PROC PRINT DATA=TEMP;
```

Because TOTAL.TAXES must be in order by STATE, PROC SUMMARY could have used a BY statement instead of a CLASS statement. In that case, the NWAY option could be eliminated.

CHAPTER 7

Data Manipulation

Overview

Compared to other tasks such as reading or sorting data, most data manipulation runs relatively quickly. After all, if the statement X=2; were slow, nobody would use the software. As a result, most of the savings described in this chapter are relatively small. On the other hand, you can probably learn and ingrain these techniques rather quickly.

This chapter organizes the data manipulation topics into three sections, all geared toward reducing CPU time. Section 1 describes working with functions. Section 2 describes various types of loops. Section 3 describes miscellaneous data manipulation issues, such as working with arrays.

Legend for Icons

CPU time or storage space
Black stars = technique applies frequently
White stars = technique applies infrequently

3 black or white stars - maximum savings
2 black or white stars - medium savings
1 black or white star - smallest savings

Fluctuating savings
Savings can range from large to nonexistent to negative, depending on your operating system or data characteristics.

Archaic
No savings under current release.

Section 1: Functions, Good and Bad Use

Functions can save a lot of programming time. By definition, the programming logic is built into the function and has already been tested and debugged. On the other hand, functions calls can sometimes be replaced by faster code.

Replacing a Function Call in General

Beware of calling the same function multiple times. Instead, create a new variable to hold the output from the function call. In this example, the INDEX function searches through a character string, looking for the first occurrence of a shorter string:

```
DATA DOCTORS MEDICAL;
SET ALLNAMES;
IF INDEX(NAME, 'M.D.') THEN OUTPUT DOCTORS;
IF INDEX(NAME, 'M.D.') OR INDEX(NAME, 'R.N.')
   THEN OUTPUT MEDICAL;

DATA DOCTORS MEDICAL;
SET ALLNAMES;
FOUND = INDEX(NAME, 'M.D.');
IF FOUND THEN OUTPUT DOCTORS;
IF FOUND OR INDEX(NAME, 'R.N.')
   THEN OUTPUT MEDICAL;
DROP FOUND;
```

Even though the second DATA step creates an extra variable, it runs faster. The extra variable allows the DATA step to cut out a function call.

The savings in CPU time vary, depending on which function gets eliminated from the program. But the general process – execute the function once and retain the results – should always run faster.

When the raw data contain the proper values, read from the raw data instead of calculating via a function. The longer-running program uses function calls to create MON and YR:

```
DATA DATES;
INFILE RAW;
INPUT DATE MMDDYY8.;
MON = MONTH(DATE);
YR  = YEAR(DATE);
```

For the INPUT statement to read DATE, the raw data must contain the month and a two-digit year. In that case, read the raw data instead of adding functions:

```
DATA DATES;
INFILE RAW;
INPUT DATE MMDDYY8. @1 MON 2.  @7 YEAR 2.;
```

If you want a four-digit year, you can always modify YEAR.

Let's consider a few examples where you can replace a function with faster code. Sometimes the new code is simple, but sometimes you must trade off writing a more complex program for the savings in CPU time.

Replacing SUBSTR Calls

The SUBSTR function copies a portion of a character string:

```
DATA NEW;
SET OLD;
LENGTH FIRST3 $ 3;
FIRST3 = SUBSTR(ORIGINAL, 1, 3);

DATA NEW;
SET OLD;
LENGTH FIRST3 $ 3;
FIRST3 = ORIGINAL;
```

The assignment statement in the second DATA step runs at least three times as fast as the SUBSTR function. On some operating systems, the difference was even more pronounced. Chapter 10 presents the details of all tests.

Even the first DATA step requires the LENGTH statement. Without it, SUBSTR would have assigned to FIRST3 the same length as ORIGINAL even though the variable really needs only three bytes. Chapter 8, "Storage Space," explains the reasons for this odd length.

Once the program applies a LENGTH statement, it no longer needs the SUBSTR function. The simple assignment statement in the second DATA step copies as much of ORIGINAL as will fit into FIRST3. Similar considerations apply when comparing the first portion of a variable to a character string. This type of comparison should not call a function:

```
IF SUBSTR(LASTNAME,1,1) = 'V';
```

Instead, add the colon modifier to the comparison:

```
IF LASTNAME =: 'V';
```

Normally, when the software compares strings of unequal length, it appends blanks to the end of the shorter string. The colon forces the software to take the opposite approach, truncating the longer string so it can make the comparison based on the length of the shorter string. The colon can be appended to all character comparison operators, not just the equal sign. In all cases, appending blanks or truncating a string does not affect the value of any variables involved; appending or truncating applies only to the process of making a comparison.

Replacing TODAY() Calls

The TODAY() function returns the current date. The INTNX function returns the first day of a time period. In the DATA steps below, the INTNX function computes the beginning and ending day of the month, based on the current date:

```
DATA NEW;
SET OLD;
D_BEGIN = INTNX('MONTH', TODAY(), 0);
D_END   = INTNX('MONTH', TODAY(), 1) - 1;

DATA NEW;
SET OLD;
DATE = TODAY();
D_BEGIN = INTNX('MONTH', DATE, 0);
D_END   = INTNX('MONTH', DATE, 1) - 1;
DROP DATE;
```

By creating DATE, the second DATA step executes the TODAY() function once instead of twice per observation. An even faster program would call each function once for the entire DATA step instead of once per observation. Because DATE, D_BEGIN, and D_END take on constant values across all observations, compute them once and RETAIN the results:

```
DATA NEW;
SET OLD;
IF _N_=1 THEN DO;
   DATE=TODAY();
   D_BEGIN = INTNX('MONTH', DATE, 0);
   D_END   = INTNX('MONTH', DATE, 1) - 1;
   RETAIN DATE D_BEGIN D_END;
   DROP DATE;
END;
```

Although the program must compare _N_ vs. 1 on each observation, that comparison runs faster than calling functions on each. The CPU time for the TODAY() function approximately equals the CPU time for reading a variable from raw data (with some variation across operating systems). By using the automatic macro variable &SYSDATE,

the program can go one step further, cutting out the TODAY()
function:

```
DATA NEW;
SET OLD;
IF _N_=1 THEN DO;
   D_BEGIN = INTNX('MONTH', "&SYSDATE"D, 0);
   D_END   = INTNX('MONTH', "&SYSDATE"D, 1) - 1;
   RETAIN D_BEGIN D_END;
END;
```

&SYSDATE is a seven-character expression for today's date in DATE7
form such as 01APR99. To avoid problems with two-digit years and the
YEARCUTOFF option, Version 7 adds &SYSDATE9 to the list of automatic
macro variables. It holds a nine-character date that includes a four-
digit year instead of a two-digit year. Once again, the bottom line is
that you should examine whether functions are producing the same
output on every observation. Call such functions once and retain their
output.

Replacing LAG Calls

Next, let's eliminate calls to the LAG function. The LAG function
retrieves a variable's value from an earlier observation. Most applications that call the LAG function intend to retrieve from the immediately preceding observation. This function really performs a tricky
operation, however. It retrieves a variable's value from the most
recent observation on which the LAG function executed. This may or
may not be the immediately preceding observation. If your DATA step
deletes observations or conditionally calls the LAG function with an IF
THEN statement, LAG can retrieve from an earlier observation than the
immediately preceding one.

Replacing a call to the LAG function is easy. Both the second and
third DATA steps are capable of replacing the first:

```
DATA NEW;
SET OLD;
Y = LAG(X);
```

```
DATA NEW;
SET OLD;
OUTPUT;
Y = X;
RETAIN Y;

DATA NEW;
LENGTH X Y $ 3;
Y = X;
SET OLD;
```

Without a LENGTH statement, the assignment statement in the third DATA step would have defined X and Y as numeric. So the DATA step needs a LENGTH statement whenever those variables should be character. Notice how the final DATA step takes advantage of the fact that variables read from a SAS data set are automatically retained. The value of X from the previous observation is retained and available to the assignment statement that creates Y. As Chapter 10 reports in more detail, removing LAG function calls reduces CPU time consistently for two operating systems (MVS and UNIX), reducing by 7% the CPU time for an entire DATA step.

Replacing PUT Calls

The next function we will consider is PUT (similar considerations apply to the INPUT function). Typically, programs call the PUT function to convert from numeric to character and the INPUT function to convert from character to numeric. For example, these statements create a character variable CHARVAR with a value of 125:

```
VALUE = 125;
CHARVAR = PUT(VALUE, 3.);
```

You could, alternatively, allow the software to control the conversion:

```
VALUE = 125;
LENGTH CHARVAR $ 3;
CHARVAR = VALUE;
```

Under some releases of the software the PUT function is faster; under others the assignment statement is faster. However, the assignment statement always generates a message in the SAS Log about numeric to character conversion. This message takes some of your time later to examine and confirm that it is benign. Recommendation: stick with the PUT (or INPUT) function, even if an assignment statement would run faster.

Section 2: Loops

SAS software supports a variety of loops, which provide opportunities to reduce CPU time.

Control DO Loops

Each iteration of a DO loop performs work. Consider this loop, for example:

```
DO YEAR = START TO FINISH;
   BALANCE = BALANCE * (1 + INTEREST);
END;
```

In addition to calculating BALANCE, each iteration of the loop carries out a series of steps at the END statement: add 1 to the current value of YEAR, compare YEAR to FINISH, and then, if YEAR is still within range, return control to the DO statement. Cut out this work by ending your loops as soon as possible.

Let's consider an opportunity to reduce the number of DO loop iterations. Here is the situation:

- The data set VALIDIDS contains one observation with variables ID1 through ID50, which are guaranteed to hold 50 different values.

- The data set OLD contains ID, which determines whether an observation should be processed or deleted. The program should delete observations where ID fails to match any of the 50 values in VALIDIDS.

This initial program works:

```
DATA NEW;
IF _N_=1 THEN SET VALIDIDS;
ARRAY IDS {50} ID1-ID50;
SET OLD;
DO I=1 TO 50;
   IF ID = IDS{I} THEN OUTPUT;
END;
```

However, this program performs extra work. If ID happens to match ID24, there is no need to check the remaining 26 array elements. After all, they are guaranteed in this example to be different. However, the DO loop continues through all 50 elements. The program should exit the DO loop as soon as a match is found. For example:

```
DO I=1 TO 50;
   IF ID = IDS{I} THEN RETURN;
END;
DELETE;
```

Or, if the DATA step should continue to process the desired observations:

```
DO I=1 TO 50;
   IF ID = IDS{I} THEN GOTO NEXT;
END;
DELETE;
NEXT:  A = B + C;
```

The LEAVE statement will also exit a DO loop, continuing with the statement following the END statement.

When nesting DO loops, changing the order of the nesting can reduce the number of loops. Consider these three variations on nested loops.

Variation #1:

```
DO CUSTOMER = 1 TO 100;
   DO PRICE = 5 TO 25 BY 5;
      DO QUANTITY = 1 TO 3;
         TOTCOST = PRICE * QUANTITY;
         OUTPUT;
      END;
   END;
END;
```

Variation #2:

```
DO QUANTITY = 1 TO 3;
   DO PRICE = 5 TO 25 BY 5;
      DO CUSTOMER = 1 TO 100;
         TOTCOST = PRICE * QUANTITY;
         OUTPUT;
      END;
   END;
END;
```

Variation #3:

```
DO QUANTITY = 1 TO 3;
   DO PRICE = 5 TO 25 BY 5;
      TOTCOST = PRICE * QUANTITY;
      DO CUSTOMER = 1 TO 100;
         OUTPUT;
      END;
   END;
END;
```

Switching from variation #1 to variation #2 reveals that the program is performing extra work. The computation PRICE * QUANTITY is constant for all 100 values of CUSTOMER. Therefore, variation #3 moves that computation outside the inner loop, making 15 calculations instead of 1,500.

Now comes the surprising part. While switching from variation #2 to variation #3 does run faster, greater savings result from switching from variation #1 to variation #2! Where do the savings come from? Let's examine how much work each variation performs.

Variation	Inner Loops	Middle Loops	Outer Loops	Total Loops	Assignment Statements
#1	1,500	500	100	2,100	1,500
#2	1,500	15	3	1,518	1,500
#3	1,500	15	3	1,518	15

When nesting DO loops, you can reduce the number of loops by placing the variable that changes the most in the innermost loop (and the variable that changes the least in the outermost loop). In some applications, the programming logic prevents you from changing the order of nested DO loops. In others, you can easily change the order. Consider, for example, this three-dimensional array:

```
ARRAY THREED {2, 150, 10} V1-V3000;
```

How would you change all elements to zero? Variation #1 below executes 4,650 loops, while variation #2 executes 3,022 loops:

Variation #1:

```
DO J = 1 TO 150;
    DO K = 1 TO 10;
        DO I = 1 TO 2;
            THREED{I, J, K} = 0;
        END;
    END;
END;
```

Variation #2:

```
DO I = 1 TO 2;
    DO K = 1 TO 10;
        DO J = 1 TO 150;
            THREED{I, J, K} = 0;
        END;
    END;
END;
```

Checking for Missing Values

On the surface, this statement does not appear to contain a loop:

```
TOTAL = MISSING + A + B + C + D;
```

Beneath the surface, however, a loop emerges when one if the incoming variables is missing. If D were missing, the software would add the other four variables, and then call a routine to track that missing values were generated by performing a mathematical operation on missing values. However, if MISSING were missing instead of D,

the software would call that routine four times (once for each attempt to add) instead of once. To circumvent this loop, you can try two techniques. If you know that one of the variables is frequently missing, condition on it being nonmissing:

```
IF MISSING > . THEN TOTAL = MISSING + A + B + C + D;
```

Alternatively, place that variable at the end of the expression. If MISSING were the variable that is usually missing:

```
TOTAL = A + B + C + D + MISSING;
```

This statement runs about three times as fast as the original, for observations where MISSING is missing.

The DATA Step Loop

The DATA step loop is another type of loop that can affect efficiency. In a simple DATA step like the one below, the SET and assignment statements execute for each observation, not just once for the entire DATA step:

```
DATA NEW;
SET OLD;
CUPS = 2*PINTS + 4*QUARTS;
```

The software automatically loops through the DATA step statements until the SET statement runs out of observations. Let's consider two sample programs that perform extra work because of this looping aspect to the DATA step. First:

```
DATA AVERAGES;
SET SALES;
BY CUSTOMER;
N + 1;
TOTAL + PRICE * (1 + TAXRATE);
AVG = TOTAL / N;
IF LAST.CUSTOMER THEN DO;
   OUTPUT;
   TOTAL = 0;
   N = 0;
END;
```

Because of the DATA step loop, the program calculates AVG on each observation rather than calculating AVG after reading all the data. Because the DATA step discards all but the final observation for each

CUSTOMER, the calculation of AVG should appear inside the DO group instead of outside.

The second example requires a little more thought:

```
DATA _NULL_;
SET SALES (KEEP=STATE) END=LASTONE;
BY STATE;
IF FIRST.STATE THEN N + 1;
IF LASTONE;
CALL SYMPUT('STATES', PUT(N, 2.));
```

The program accurately counts the number of states in the data and captures that as a macro variable. However, the program performs extra work by checking for end-of-file on each observation. Logically, the final observation in the data set must be the last one for some state. Therefore, replace the highlighted statements with:

```
IF LAST.STATE;
N + 1;
IF LASTONE;
```

Now the check for end-of-file takes place on just those observations that are the last one for a state.

Section 3: Miscellaneous Data Manipulation Techniques

Tips for Mathematics

A few tips will signifcantly speed up computations:

- multiply, don't divide

- check for division by zero

- group numeric constants.

Let's look at a brief example of each.

Multiplication runs faster than division. To streamline your program, you could switch from the first statement to the second:

```
J = I / 2;
J = I * 0.5;
```

While the percentage savings can be large (on the order of 25%), the absolute savings are small.

Next, consider the possibility of division by zero:

```
QUOTIENT = NUMER / DENOM;
```

On observations where DENOM is zero, QUOTIENT ends up missing and the software reports in the SAS Log that division by zero occurred. If that is a remotely possible event, consider switching to this statement instead:

```
IF DENOM THEN QUOTIENT = NUMER / DENOM;
```

The condition IF DENOM is false whenever DENOM is zero or missing. When DENOM is zero, the original statement uses over 400 times the CPU time as the replacement!

Finally, the software looks for numeric constants and preprocesses a group of constants into one. For example, the first statement is more efficient than the second:

```
TOTAL = 1 + 8 + AMOUNT;
TOTAL = 1 + AMOUNT + 8;
```

In the first statement, the software adds 1 + 8 once for the entire DATA step, and then treats the statement as if it were

```
TOTAL = 9 + AMOUNT;
```

Adding twice takes about twice as long as adding once. So where possible, group numeric constants in a mathematical expression.

Reduce the Number of Comparisons

Each comparison utilizes a small amount of CPU time:

```
IF X=2 THEN DO;
```

Although comparing X to 2 requires a tiny amount of CPU time, DATA steps often make dozens of comparisons per observation. We'll consider methods to reduce that number.

When making a series of comparisons, use ELSE to cut down the number of comparisons. This set of statements examines NAME three times on each observation:

```
LENGTH CATEGORY $ 8;
IF NAME='WILLIAM' THEN CATEGORY='HUSBAND';
IF NAME='HILLARY' THEN CATEGORY='WIFE';
IF NAME='CHELSEA' THEN CATEGORY='DAUGHTER';
```

A faster set would examine NAME a maximum of three times per observation:

```
LENGTH CATEGORY $ 8;
IF      NAME='WILLIAM' THEN CATEGORY='HUSBAND';
ELSE IF NAME='HILLARY' THEN CATEGORY='WIFE';
ELSE IF NAME='CHELSEA' THEN CATEGORY='DAUGHTER';
```

This new set examines NAME three times on some observations (when NAME is CHELSEA), but just once or twice per observation on others (when NAME is WILLIAM or HILLARY). In addition, if you know the distribution of NAMEs within the data, order the statements from highest to lowest frequency of occurrence. For example, the second set above runs fastest when WILLIAM is the most frequently occurring value for NAME, followed by HILLARY and then CHELSEA. If HILLARY were the most frequently occurring value, followed by CHELSEA and then WILLIAM, the fastest order would be:

```
LENGTH CATEGORY $ 8;
IF        NAME='HILLARY' THEN CATEGORY='WIFE';
ELSE IF NAME='CHELSEA' THEN CATEGORY='DAUGHTER';
ELSE IF NAME='WILLIAM' THEN CATEGORY='HUSBAND';
```

With a series of numeric ranges, beware of making extra comparisons:

```
LENGTH STATUS $ 14;
IF        (      AGE <  1) THEN STATUS='BABY';
ELSE IF ( 1 <= AGE <  2) THEN STATUS='INFANT';
ELSE IF ( 2 <= AGE <  4) THEN STATUS='TODDLER';
ELSE IF ( 4 <= AGE < 13) THEN STATUS='CHILD';
ELSE IF (13 <= AGE < 20) THEN STATUS='TEENAGER';
ELSE IF (20 <= AGE < 30) THEN STATUS='YOUNG ADULT';
ELSE IF (30 <= AGE < 45) THEN STATUS='ADULT';
ELSE IF (45 <= AGE < 65) THEN STATUS='MIDDLE AGED';
ELSE IF (65 <= AGE      ) THEN STATUS='SENIOR CITIZEN';
```

An ELSE statement executes when the previous IF THEN condition was examined and found to be false. For example, the first ELSE statement executes only when the comparison AGE < 1 is false. By extending this idea to all the ELSE statements, the program can skip half of the evaluations:

```
LENGTH STATUS $ 14;
    IF        AGE <  1 THEN STATUS='BABY';
    ELSE IF AGE <  2 THEN STATUS='INFANT';
    ELSE IF AGE <  4 THEN STATUS='TODDLER';
    ELSE IF AGE < 13 THEN STATUS='CHILD';
    ELSE IF AGE < 20 THEN STATUS='TEENAGER';
    ELSE IF AGE < 30 THEN STATUS='YOUNG ADULT';
    ELSE IF AGE < 45 THEN STATUS='ADULT';
    ELSE IF AGE < 65 THEN STATUS='MIDDLE AGED';
    ELSE                 STATUS='SENIOR CITIZEN';
```

A series of IF/THEN/ELSE statements runs just as fast as a SELECT statement. You can switch (as a matter of style, not as a matter of efficiency) to:

```
LENGTH STATUS $ 14;
SELECT;
   WHEN (AGE <  1) STATUS='BABY';
   WHEN (AGE <  2) STATUS='INFANT';
   WHEN (AGE <  4) STATUS='TODDLER';
   WHEN (AGE < 13) STATUS='CHILD';
   WHEN (AGE < 20) STATUS='TEENAGER';
   WHEN (AGE < 30) STATUS='YOUNG ADULT';
   WHEN (AGE < 45) STATUS='ADULT';
   WHEN (AGE < 65) STATUS='MIDDLE AGED';
   OTHERWISE       STATUS='SENIOR CITIZEN';
END;
```

Surprisingly, however, the alternative form, SELECT(*value*);, takes longer.

With a large number of categories, consider adding a FORMAT. If this format existed, then a DATA step could use it to calculate STATUS in a single statement:

```
PROC FORMAT;
VALUE STATUS    . - .Z,
             LOW - <  1 = 'BABY'
               1 - <  2 = 'INFANT'
               2 - <  4 = 'TODDLER'
               4 - < 13 = 'CHILD'
              13 - < 20 = 'TEENAGER'
              20 - < 30 = 'YOUNG ADULT'
              30 - < 45 = 'ADULT'
              45 - < 65 = 'MIDDLE AGED'
              OTHER      = 'SENIOR CITIZEN';
```

The DATA step would use the PUT function:

```
STATUS = PUT(AGE, STATUS.);
```

Because the PUT function performs a binary search on the format table, this approach becomes (relatively) faster as the number of recoding categories grows. The speed of the PUT function varies from one release of the software to the next.

Some statements evaluate compound conditions:

```
IF X=5 AND Y=10 THEN DO;
```

The software first evaluates X=5. Only when that comparison is true does it evaluate Y=10. The following statement attempts to avoid evaluating Y=10 unless X is 5:

```
IF X=5 THEN IF Y=10 THEN DO;
```

Such statements are unnecessary because the software automatically stops evaluating either statement when X is not 5. Therefore, with multiple conditions joined by AND, your program runs slightly faster when the condition that is usually false appears first.

The IN operator checks a series of values, but is slower than a series of OR conditions:

```
IF SYMBOL IN ('GE', 'GM', 'F');
IF SYMBOL = 'GE' OR SYMBOL='GM' OR SYMBOL='F';
```

However, the IN operator may be easier to read, reducing program maintenance time.

The WHERE statement attempts to combine multiple conditions:

```
IF X=5 OR Y=10 OR X=10 THEN DO;
```

Automatically, the software evaluates X=5 first, X=10 second, and Y=10 last. The statement behaves as if it were:

```
IF X IN (5, 10) OR Y=10 THEN DO;
```

You can run your own tests to check out other possibilities. For example:

- Will the software always evaluate multiple conditions from left to right?

- Does the WHERE statement evaluate multiple conditions differently than an IF or IF THEN statement?

Chapter 10 presents a limited set of test results for evaluating multiple conditions.

Finally, let's consider how programs evaluate numeric constants as true or false. While these statements produce equivalent results, the first runs slightly faster than the second:

```
IF FIRST.STATE    THEN DO;
IF FIRST.STATE=1 THEN DO;
```

When the software evaluates numeric constants, it takes zero and missing values to be false and any other value (including negative numbers) to be true. Therefore, the first statement is true when FIRST.STATE=1 and false when FIRST.STATE=0. When the software evaluates the second statement, behind the scenes it replaces true comparisons with a 1 and false comparisons with a 0. Next, it evaluates these numeric constants as being true or false. So the second statement performs an extra step. The same considerations apply to IN= and END= variables.

Arrays: Myths and Realities

Many techniques throughout the literature claim to produce small gains in efficiency when using arrays. The list includes:

- Use arrays only when they are needed.

- Refer to variable names (or constant values) instead of array elements.

- Use a temporary array for a set of numeric constants.

- Don't mix retained and nonretained variables in the same array.

- Assign the same length to all variables in an array.

- Don't let a variable appear in multiple arrays.

- Use 0 (not 1) as the lower bound for each dimension in the array.

Instead of working through detailed examples for potentially small savings, just note that the first two steps provide the largest savings. The others provide very little savings, including some that are so small they are impossible to measure. The last item, using 0 as the lower bound to an array, saves having to execute a single instruction which, for all intents and purposes, is not measurable. Instead, change the lower bound for practical considerations, such as:

```
ARRAY YEARS {9} FY1991-FY1999;
PERCENT = YEARS{YR - 1990} / TOTAL * 100;

ARRAY YEARS {1991:1999} FY1991-FY1999;
PERCENT = YEARS{YR} / TOTAL * 100;
```

By switching from the top to the bottom set of statements, the program becomes easier to read, and it eliminates substracting 1990 from YR.

Compilation vs. Execution

When running a DATA step, assignment statements execute on each observation. Even the simplest assignment statement like this one uses some CPU time on every observation:

```
X = 25;
```

Because X is constant, a faster statement would assign X the value 25 once and allow that value to sit, never changing:

```
RETAIN X 25;
```

Along similar lines, these three statements are listed from slowest to fastest:

```
DATE = TODAY();

IF _N_=1 THEN DATE=TODAY();
RETAIN DATE;

RETAIN DATE "&SYSDATE"D;
```

The RETAIN statement can assign an initial value, and then hold on to that value from one observation to the next. As a result, you can speed up a DATA step by retaining all variables (subject to the logic requirements of the program). When moving from one observation to the next, the software performs less work on retained variables. Nonretained variables must be set to missing, but retained variables require no changes.

A Small Workshop

Problem #1

How many comparisons are made by the following code? Answer for two cases: the incoming data contain 100 males and 200 females, and the incoming data contain 100 females and 200 males.

```
IF GENDER='F' THEN TYPE='FEMALE';
ELSE IF GENDER='M' THEN TYPE='MALE';
```

Problem #2

Improve the efficiency of these two sets of code:

```
IF RATE <= 0 THEN INTEREST = 'INVALID';
IF RATE > 10 THEN INTEREST = 'HIGH';
IF 0 < RATE <= 8 THEN INTEREST = 'LOW';
IF 8 < RATE <= 10 THEN INTEREST = 'MEDIUM';

INPUT DATE $6.;   /* YYMMDD form */
IF SUBSTR(DATE,1,2) > '80' THEN NEWDATE = '18' || DATE;
IF SUBSTR(DATE,1,2) < '10' THEN NEWDATE = '20' || DATE;
IF SUBSTR(DATE,1,2) <= '80' AND SUBSTR(DATE,1,2) >= '10'
   THEN NEWDATE = '19' || DATE;
```

Problem #3

Improve the efficiency of this code, while still using arrays:

```
DATA PERCENTS;
SET ALLYEAR;
ARRAY MONTHS {12} MONTH1-MONTH12;
ARRAY PCTS   {12} P1-P12;
IF YEAREND='YES' THEN DO I=1 TO 12;
   PCTS{I} = MONTHS{I} / SUM(OF MONTH1-MONTH12);
END;
```

Workshop Solutions

Problem #1

400 and 500. The number of males gets compared twice, to both F and M. The number of females gets compared once, to F.

Problem #2

For the first set of code, use ELSE for mutually exclusive conditions:

```
IF        RATE <=  0 THEN INTEREST = 'INVALID';
ELSE IF RATE <=  8 THEN INTEREST = 'LOW';
ELSE IF RATE <=10 THEN INTEREST = 'MEDIUM';
ELSE                      INTEREST = 'HIGH';
```

For the second set, get rid of the SUBSTR calls, and use ELSE for mutually exclusive conditions:

```
INPUT DATE $6.;   /* YYMMDD form */
IF        DATE >:  '80' THEN NEWDATE = '18' || DATE;
ELSE IF DATE >=: '10' THEN NEWDATE = '19' || DATE;
ELSE                        NEWDATE = '20' || DATE;
```

Besides being clearer and running faster, this code gets the right answer while the original code got the wrong answer! In the original, years greater than 80 receive a prefix of 18 and therefore receive a second prefix of 19.

Problem #3

Compute the sum of MONTH1 through MONTH12 once per observation. As a minor consideration, remember that multiplication is faster than division.

```
DATA PERCENTS;
SET ALLYEAR;
ARRAY MONTHS {12} MONTH1-MONTH12;
ARRAY PCTS   {12} P1-P12;
IF YEAREND='YES' THEN DO;
   MULTBY = 1 / SUM(OF MONTH1-MONTH12);
   DO I=1 TO 12;
      PCTS{I} = MONTHS{I} * MULTBY;
   END;
END;
```

CHAPTER 8

Storage Space

Overview

Many techniques that save CPU time also reduce storage space requirements. Earlier chapters have already covered these related topics:

- Save only needed variables (Chapter 2)

- Read several raw data files in one step (Chapter 2)

- Test programs on a sample of the data to temporarily save space (Chapter 2)

Legend for Icons

CPU time or storage space
Black stars = technique applies frequently
White stars = technique applies infrequently

3 black or white stars - maximum savings
2 black or white stars - medium savings
1 black or white star - smallest savings

Fluctuating savings
Savings can range from large to nonexistent to negative, depending on your operating system or data characteristics.

Archaic
No savings under current release.

- Use WHERE in a procedure instead of creating a subset (Chapter 2)

- Perform all data manipulation in one DATA step (Chapter 4)

- Use _NULL_ as a data set name (Chapter 3)

- Skip unnecessary SORTs (Chapter 5).

On the other hand, some CPU time savers use more space:

- Storing permanent SAS data sets

- Creating an index

- Storing summary data sets.

For this last group of items, you will have to weigh which resource is most valuable (the CPU time, the storage space, or the programming time required to learn and implement new techniques). There is no definitive right or wrong time to use these techniques because the relative cost and availability of resources vary from site to site.

This chapter does not review the above topics. Instead, it examines additional tools that save storage space. Section 1 explains how to reduce the size of an individual SAS data set. Section 2 tackles reducing the amount of space used across all data sets within a SAS data library. Finally, Section 3 looks at other storage forms, such as tape, views, and formats.

Section 1: Shrinking an Individual SAS Data Set

Two types of techniques can reduce the size of a SAS data set:

- controlling the lengths of individual variables

- compressing the data set.

This section examines both topics.

Lengths of Variables: Common Methods

If you want to control the lengths of variables, it is important to understand how the software selects those lengths. As the software compiles your DATA step statements, it assigns a length to each variable. In general, the first mention of a variable determines its length. However, note that:

- A few statements and options do not supply enough information to assign attributes such as length or type. For example, a KEEP statement (or the equivalent KEEP= data set option in the DATA statement) does not assign length. Also, a RETAIN statement assigns length to only those variables that receive an initial value. For example, assume that the following statement represents the first mention of V1, V2, and V3:

```
RETAIN V1 5 V2 'BOB' V3;
```

  V1 receives a length of 8, the default for a numeric variable. V2 receives a length of 3, the number of characters in BOB. V3 receives neither length nor type attributes.

- A subsequent LENGTH statement can truncate numeric variables in the output data set. The same does not apply to character variables, however. Attempting to reset the length of a character variable with a LENGTH statement generates an error.

For character variables the first mention of a variable determines its length. Usually this first mention occurs in one of four ways:

- A LENGTH statement. Here, the program explicitly assigns a length to the variable(s):

```
LENGTH  LNAME $ 15  FNAME $ 10;
```

- A SET statement. The SET statement automatically assigns a length to each variable from the incoming data set, assigning the same length in the new data set as the variable had in the original.

- The INPUT statement. When reading with column input, the length is the number of characters read. When reading with list input, the length is always 8. When reading with formatted input, the length is often the length of the informat, but can vary. For example, consider this INPUT statement:

```
INPUT   NAME $            /* length =  8 */
        RANK1 $ 21-30     /* length = 10 */
        @21 RANK2 $10.    /* length = 10 */
        @31 H $HEX2.;     /* length =  1 */
```

List input assigns NAME a length of 8. Column input reads 10 characters, and therefore assigns RANK1 a length of 10. Formatted input also reads 10 characters, assigning RANK2 a length of 10. However, formatted input assigns H a length of 1 because the software determines that the $HEX2 informat requires only one character of storage space to hold the final value.

- A programming statement. FNAME, TYPE, and DEGREE below would receive a length of 4 if the sample code created the variable:

```
FNAME='FRED';

IF GENDER='M' THEN TYPE='MALE';
ELSE TYPE='FEMALE';

DO UNTIL (DEGREE='M.D.');
```

Of these four methods, only the first two apply to numeric variables. Numeric variables receive a length of 8 when created by an INPUT statement or assignment statement, except in special cases noted below.

Control Variable Lengths: Standard Situations

Obviously, using shorter lengths for variables decreases the size of a SAS data set. As a preliminary step, you must have examined your data and determined the lengths you will need.

Use the programming techniques listed above to control lengths of variables. In the sample code above, TYPE takes on a length of 4 because the earliest mention of it sets it equal to MALE, which is four characters long. Any of three methods would assign TYPE the necessary length of 6:

```
LENGTH TYPE $ 6;   /* before any IF THEN statements */

IF GENDER='M' THEN TYPE='MALE  ';   /* embed blanks */
ELSE TYPE='FEMALE';

IF GENDER='F' THEN TYPE='FEMALE';   /* FEMALE first */
ELSE TYPE='MALE';
```

Statements that read data (INPUT, SET, MERGE, and UPDATE) assign a length to each new variable they read, but do not change the length of an existing variable. Therefore, if you desire a different length, place the length statement BEFORE the "read" statement. In that way, the read statement treats the variable as an existing variable with a predefined length.

Numeric variables always use a length of 8, unless you specify a different length using either:

- a LENGTH statement

- the DEFAULT= option in the INFILE statement.

If you do use a LENGTH statement for numeric variables, you can place it anywhere in the DATA step. All numerics have a length of 8 in the Program Data Vector. The truncation to a smaller length occurs as observations leave the Program Data Vector.

What does a length of 8 mean? SAS software stores numbers in a binary floating point form, meaning that the first portion of the variable holds information about the sign (positive or negative) and the location of the decimal point. The remainder of the variable holds a binary integer. Different operating systems use somewhat different formats for the initial portion of the variable. The easiest example to use is MVS, where one byte exactly stores this information. Bytes 2 through 8 store a binary integer. In that case, with a length of 2, the maximum integer value that can be accurately stored is 255. This number is 2^8-1. Because the second byte contains 8 bits, any of which can be turned on or off, there are 2^8 combinations which represent the integers 0 through 255. With a length of 3 there would be 2^{16} combinations representing integers from 0 through $2^{16}-1$ or 65,535. Under other operating systems, the principal remains the same. However, other operating systems use more than one byte to hold the sign and decimal point information, and therefore can store different maximum integer values for a given variable length.

What does all this mean for setting the lengths of numeric variables? Under normal circumstances, unless variables take on integer values, you lose precision by truncating the lengths of numeric variables. Change the length under these conditions only:

- The variable takes on integer values only (and you know what the largest possible value is going to be).

- You don't care about losing precision.

Date variables are a good candidate for truncating length. They take on integer values only, and you frequently know what the largest or smallest value will be. Under MVS, a length of 3 handles current dates because 65,535 represents over 150 years both forward and backward from January 1, 1960. A length of 4 would cover the entire date scale and then some ($2^{24}-1 = 16,777,215$). Under other operating systems, a length of 4 might be necessary, even for current dates. Check the companion guide for your operating system to discover the relationship between the largest integer value stored and the length of a numeric variable.

The stock market represents an interesting but less common opportunity to shorten numeric variables. Individual stock prices take on fractional values like 35½ or 112¼. While these numbers are not integers, a binary system can represent these values exactly.

Base 10 Representation	Binary Representation
4	100
2	10
1	1
0.5	0.1
0.25	0.01
35.5	100011.1
112.25	1110000.01

Two bytes of a binary integer can accurately store fractional values as small as one sixty-fourth and as large as 1,023 and sixty-three sixty-fourths.

Consider storing dollar and cents amounts as integers, storing data values as the number of pennies. By storing integers, this strategy preserves precision, while allowing you to truncate variable lengths. In addition, you don't have to convert the numbers for printing purposes! This format will insert the decimal points for printing nonnegative integers:

```
PICTURE INSERTD
        0-HIGH = ' 0,000,001.11' (prefix='$' mult=1);
```

Choose the Shorter Variable Type

Variables that take on single digit values require a length of 1 as a character variable but a longer length as a numeric. If the variable truly represents a category rather than a quantity, consider storing the information as a character variable. Five-digit ZIP codes are categorical by nature. You will never add ZIP codes together or multiply ZIP code by some other variable. Although ZIP codes require a length of 5 as a character variable, you can store ZIP codes with a length of 4 as numeric. On the other hand, you must remember to format the variable with the Z5 format if you want to print leading zeros in your reports. Nine-digit zip codes and social security numbers require at least nine bytes of storage as character variables. As numeric integers, a length of five is more than ample. On the other hand, you may need to create picture formats to print these variables in easy to read forms.

Functions and Lengths of Variables

Pay attention to the lengths of character variables whenever a function creates the variables. Consider these examples where functions create new variables:

```
USERID  = SUBSTR(DATASET, 1, 7);
LNAME   = TRIM(LASTNAME);
NOBLANKS = COMPRESS(STRING);
```

In every case, the new variable receives the same length as the incoming variable. In a way, this result makes sense. In the case of the SUBSTR function, the third parameter does not have to be hard-coded. It might be an expression based on numeric variables in the data set. In that case, the software "knows" that USERID will be some portion of DATASET. Conceivably, the entire value of DATASET might be selected. Therefore the software must assign USERID the same length as DATASET. Perhaps future releases of the software will examine the third argument to SUBSTR to see whether it contains an expression or a hard-coded value. Certainly, shorter lengths would be possible when the third parameter is hard-coded.

The other functions, TRIM and COMPRESS, both have valid reasons to assign the new variable the same length as the old. LASTNAME might contain both short and long values, including some that have no trailing blanks. The trimmed variable might require the same length as the old. The same idea applies to COMPRESS. When the incoming character string contains no blanks, the new variable will require the same length as the old.

Five other functions also assign the new variable the same length as the original string:

```
LEFT, RIGHT, REVERSE, TRANSLATE, UPCASE
```

Three functions are even more dangerous in their liberal use of space:

```
ONEWORD  = SCAN(STRING, N);
NEWVALUE = SYMGET('MACROVAR');
LONGLINE = REPEAT('LINE', 8);
```

For these three functions, the new variables receive lengths of 200, the Version 6 maximum for a character variable!

In every case, a simple solution exists. If you know that a variable requires less space than the function assigns, define that variable with a LENGTH statement earlier in the DATA step.

Read Dates Intelligently

Read dates as one variable, not three:

```
INPUT MONTH 21-22 DAY 24-25 YEAR 27-28;   /* too wide */

INPUT @21 DATE MMDDYY8.;                   /* better   */
```

Three numeric variables take up three times the storage space.

Under certain circumstances, you might reverse course and read dates as multiple variables. Consider these special cases:

- The data contain partial dates. In that case, DATE in the second INPUT statement would receive a missing value. However, the first INPUT statement would retain as much information as possible. In that situation, read the month, day, and year as character rather than numeric.

- The analysis requires more than one variable, such as the year as well as the date. As mentioned in Chapter 7, creating a variable with a function takes longer than reading the information from the raw data:

```
YEAR = YEAR(DATE);                              /* longer  */

INPUT @21 DATE MMDDYY8.  @27 YEAR 2.;  /* shorter */
```

Even though the function creates a four-digit year while the INPUT statement creates a two-digit year, both variables take up 8 bytes by default.

Compress SAS Data Sets

The good news is that compressing a SAS data set is easy and might save a considerable amount of storage space. Just add the proper option in the DATA statement, for example:

```
DATA NEW (COMPRESS=YES);
```

Version 7 adds two variations:

```
DATA NEW (COMPRESS=CHAR);    /* same as YES */
DATA NEW (COMPRESS=BINARY);  /* new feature */
```

The bad news is that the results are extremely variable. You may save a lot, or you may save a little. In fact, some "compressed" SAS data sets take up more space than the original, uncompressed versions! In addition, creating and accessing compressed data uses slightly more CPU time because DATA steps and procedures process uncompressed values. Finally, direct access (retrieving observations using an index or POINT= in the SET statement) is impossible with compressed data under Version 6.

How do you know if your data set should be compressed or not? Begin by understanding what compression does.

Version 6 compression works on consecutive occurrences of the same character. A series of identical characters can be stored as a combination of one instance of the character, plus an indication of how many times that character should repeat. With a lot of repetition, compression saves a considerable amount. With little repetition, compression saves little and can even add to the size of the data set. Repetition usually involves missing values, particularly when character variables contain a series of blanks. Although other characters can theoretically be compressed, in practice missing values represent the most commonly occurring repeated characters.

The SAS System will compress repeated characters across variables. Therefore, define all character variables consecutively, without interspersing numeric variables. The compression process will be more likely to find repeated instances of the same character when all character variables occupy a continuous series of bytes.

To define character variables consecutively, add a LENGTH statement early in the DATA step. For example:

```
DATA PERM.DATA (COMPRESS=YES);
LENGTH NAME $ 30 RANK $ 20;
INFILE RAW;
INPUT NAME    $  1-30
      SERIAL     31-40
      RANK    $ 41-60;
```

Here is a more generic program that reorders the variables in an existing data set:

```
PROC CONTENTS DATA=PERM.DATA (KEEP=_CHARACTER_)
              NOPRINT OUT=TEMP;

DATA _NULL_;
CALL EXECUTE('DATA PERM.DATA (COMPRESS=YES);');
IF EOF=0 THEN DO UNTIL (EOF);
   SET TEMP END=EOF;
   CALL EXECUTE ('LENGTH ' || NAME || ' $' ||
                 PUT(LENGTH, 3.) || ';');
END;
CALL EXECUTE('SET PERM.DATA; RUN;');
RUN;
```

Note that the macro processor must be turned on, to permit CALL EXECUTE to work. Also, this approach uses an extra DATA step just to reorder the variables.

Version 7 adds a new compression routine. COMPRESS=BINARY compresses repetitions of byte patterns, rather than repetition of individual bytes. Although this method requires slightly more CPU time, it gains the advantage of more effectively compressing numeric variables. For example, three consecutive numeric variables, each with a missing value, form a repetition of the same byte pattern. No matter which compression routine you select, you will achieve better results by grouping all character variables.

How much will you save? The answer depends on characteristics of your data. Try compressing your data. Notes on the log will tell you how much you are saving.

Section 2: Conserving Space within a SAS Data Library

For practical purposes, the minimum size of a SAS data library is the sum of the sizes of all the members of the library. By following the tips in Section 1 above and thus minimizing the size of each individual data set, you are reducing the minimum size of the data library. A handful of techniques can address special situations related to data libraries, particularly the WORK library. This section examines those additional techniques.

PROC COPY Stable Libraries

This program writes two copies of BIGFILE:

```
DATA NEWLIB.BIGFILE;
INFILE RAWDATA;
INPUT ID $5. (V1-V20) (2.);

PROC SORT DATA=NEWLIB.BIGFILE;
BY ID;
```

PROC SORT reads the original version of BIGFILE, sorts it, and writes out the sorted version. However, the sorted version does not overlay the original. Instead, PROC SORT writes out the sorted version to blank space on disk, and then designates the original version as being reusable disk space. Most operating systems reuse the space automatically. However, the MVS operating system can reuse this space only for data written to NEWLIB. Therefore, under the MVS operating system, NEWLIB may contain vast amounts of unused space. If subsequent programs never write to NEWLIB, the data library takes up twice as much space as it needs! In all fairness, the operating system will reuse available space as it adds data sets to NEWLIB. Therefore, the problem lessens as the data library stores more and more data sets. Each new permanent data set starts out by using available space at the beginning of the data library.

Once your data libraries are stable, you can reclaim all embedded unused space by copying the library:

```
PROC COPY IN=NEWLIB OUT=WASTENOT;
```

You gain another advantage at the same time. Any fragmented data sets get copied to one continuous block of disk space. Later programs will read the continuous versions faster than the fragmented versions.

If your data library is 100% stable, you can release unused space at the end of the library in two ways:

- Code RLSE within the SPACE parameter of the JCL.

    ```
    SPACE=(CYL, (20,5) ,RLSE)
    ```

    Unfortunately, RLSE applies to the primary extent only, not to any secondary extents. If your new data library requires 18 cylinders, you will end up with 18. However, if it requires 23 cylinders, you will end up with 25.

- Run PROC RELEASE:

```
PROC RELEASE DDNAME=WASTENOT;
```

  PROC RELEASE will also allow you to specify an amount of blank space to leave at the end of the library.

Saving on WORK Space

If your program is running out of WORK space, PROC COPY will not help. Assuming that your individual data sets are as small as possible, here are some additional tips. The first tip applies to permanent libraries as well as the WORK library.

DELETE Old Data Sets, or Reuse Data Set Names

Assume that after this DATA step the program no longer needs the data set OLD:

```
DATA NEW;
SET OLD;
TOTCUPS = 2*PINTS + 4*QUARTS;
```

You can clear OLD out of the WORK space, along with any other data sets that the program no longer needs:

```
PROC DATASETS LIBRARY=WORK;
DELETE OLD USELESS DATA;
```

As an alternative, consider reusing data set names:

```
DATA OLD;
SET OLD;
TOTCUPS = 2*PINTS + 4*QUARTS;
```

As was the case with PROC SORT, this DATA step does not overlay the original OLD data set. Instead, it writes the newer version of OLD to blank space on disk. Once the DATA step successfully completes, the software changes the word OLD to mean the latest version. The previous version becomes reusable disk space. This technique applies to other forms of the DATA step as well, whenever the original version of OLD will no longer be needed. For example, this is perfectly legal code:

```
DATA OLD;
MERGE OLD ANCIENT;
BY ID;
TOTCUPS = 2*PINTS + 4*QUARTS;
```

The DATA step writes the new version of OLD to blank space, then treats the original version as reusable space.

Assign a Length to Dropped Numeric Variables

Intuitively, lengths for dropped variables could not make any difference. But these factors combine to reduce WORK space requirements:

- The software does not copy observations directly from the Program Data Vector to the output data set(s). Instead, it accumulates observations into an output buffer in the WORK area, which later gets copied to the output data set(s).

- The output buffer contains all variables, whether kept or dropped (not including automatic variables such as _N_ and BY variables).

- In the Program Data Vector, all numeric variables have a length of 8. In the output buffer, however, their lengths get shortened by the LENGTH statement.

Assigning a minimum length to dropped numerics does not lose precision because the Program Data Vector always stores numerics with a length of 8. However, the overall space requirements shrink because the output buffer uses less space to hold these variables.

Select Low-Space Programming Techniques

A given programing objective may permit several programming approaches. For example, in this application, ALL.COWS contains the key variables COW, PINTS, and QUARTS. The program must summarize the data down to one observation per COW, adding a new variable TOTCUPS, where:

```
TOTCUPS = 2*PINTS + 4*QUARTS;
```

For the sake of simplicity, assume that the data contain no missing values. Let's start by using excessive amounts of space:

```
DATA TOTALS;
SET ALL.COWS;
CUPS = 2*PINTS + 4*QUARTS;

PROC SORT DATA=TOTALS;
BY COW;

DATA SUMS;
SET TOTALS;
BY COW;
IF FIRST.COW THEN TOTCUPS=0;
TOTCUPS + CUPS;
IF LAST.COW;
```

The first DATA step is totally unnecessary. Instead, the final DATA step could have computed CUPS. Even if PROC SORT should not alter the original data set, this code would be much faster and would reduce WORK space by limiting TOTALS to three variables:

```
PROC SORT DATA=ALL.COWS (KEEP=COW PINTS QUARTS)
          OUT=TOTALS;
BY COW;
```

Then the DATA step would require very little change:

```
DATA TOTALS;
SET TOTALS;
BY COW;
IF FIRST.COW THEN TOTCUPS=0;
TOTCUPS + (2*PINTS + 4*QUARTS);
IF LAST.COW;
```

Once this DATA step runs, the original version of TOTALS becomes reusable disk space.

As long as there are fewer than 32,767 values for COW, PROC SUMMARY can track all the subtotals at once. The following program uses the least WORK space of all:

```
PROC SUMMARY DATA=ALL.COWS NWAY;
CLASS COW;
VAR PINTS QUARTS;
OUTPUT OUT=TOTALS (KEEP=COW TOTALP TOTALQ)
         SUM=TOTALP TOTALQ;

DATA TOTALS;
SET TOTALS;
TOTCUPS = 2*TOTALP + 4*TOTALQ;
```

PROC SUMMARY eliminates the need to sort, because it stores statistics in memory for each value of COW. So there was never any need to use WORK space, except to hold small summary data sets.

Section 3: Other Forms of Storage

Store a Code, Print a Format

Obviously, MASSACHUSETTS takes up more space than MA. If your data set stores two-letter state codes, you can always print the full name by applying a format:

```
PROC FORMAT;
VALUE $STNAME 'MA'='MASSACHUSETTS'
             'RI'='RHODE ISLAND';
PROC PRINT DATA=USA;
FORMAT STATE $STNAME.;
```

In addition to saving storage space, you automatically create other benefits at the same time. Storing a code lets you choose, when generating the report, which format to use. A second format might use initial caps instead of all caps, for example. Formats also guarantee uniform translations. With the $STNAME format, MA will always translate into MASSACHUSETTS. Without a format, data entry personnel might find a few different ways to spell or abbreviate the state name. In addition, saving and retrieving a permanent format is relatively easy. When permanent formats exist, many programs and data sets can utilize the same format, guaranteeing that different reports use the same translations.

The savings can add up quickly, particularly when the codes have a drastically shorter length than the formatted values, or when many variables use a format.

Tape vs. Disk Considerations

Storing data on tape eliminates the headache of finding enough disk space. In addition, chargeback systems usually charge less per byte for data stored on tape. On the other hand, accessing your data may require more clock time. Your job may have to wait in a longer queue, until the necessary number of tape drives becomes available. Access issues exist, as well. For example, this program becomes impossible for data stored on tape:

```
DATA TAPE1.NEW;
SET TAPE1.OLD;
```

Your program cannot simultaneously read from one spot on the tape and write to another. Instead, you would need either disk space or a second tape drive:

```
DATA TAPE2.NEW;
SET TAPE1.OLD;
```

Similarly, PROC SORT cannot read and write from the same tape. This program also fails:

```
PROC SORT DATA=TAPE3.SALES;
BY MONTH;
```

Instead, the program must write the sorted version to either disk or a second tape:

```
PROC SORT DATA=TAPE3.SALES OUT=TAPE4.SALES;
BY MONTH;
```

Finally, if your program creates multiple data sets in one DATA step, you still face the limitation of writing to one spot on the tape at a time. This program is impossible:

```
DATA TAPE6.JUNE TAPE6.JULY;
SET TAPE5.ALLYEAR;
IF        MONTH=6 THEN OUTPUT TAPE6.JUNE;
ELSE IF MONTH=7 THEN OUTPUT TAPE6.JULY;
```

Instead, each output data set must be sent to disk or to a separate tape:

```
DATA TAPE6.JUNE TAPE7.JULY;
SET TAPE5.ALLYEAR;
IF        MONTH=6 THEN OUTPUT TAPE6.JUNE;
ELSE IF MONTH=7 THEN OUTPUT TAPE7.JULY;
```

Creating a View

Instead of storing data, store a view. By definition, a view is instructions on how to extract data. Here is a simple "before" program:

```
DATA MEMBERS;
INFILE PROSPECT;
INPUT @ 1 GENDER    $CHAR1.
      @ 3 LASTNAME $CHAR12.
      @17 AGE         2.
      @20 (VAR1-VAR20) (2.);

PROC SORT DATA=MEMBERS OUT=PERM.MEMBERS;
BY AGE;
```

Once PROC SORT completes, there is no need to store the data set MEMBERS. The amazing thing is this: there is no need to store a data set MEMBERS even before PROC SORT runs. Instead, create a view:

```
DATA MEMBERS / VIEW=MEMBERS;
INFILE PROSPECT;
INPUT @ 1 GENDER    $CHAR1.
      @ 3 LASTNAME $CHAR12.
      @17 AGE        2.
      @20 (VAR1-VAR20) (2.);

PROC SORT DATA=MEMBERS OUT=PERM.MEMBERS;
BY AGE;
```

By adding / VIEW=MEMBERS, the second DATA step stores a view, not data. This means it stores instructions on how to read in data, rather than the data itself. When PROC SORT uses the view, the software reads observations from raw data into the sort workspace. There was never a need to store MEMBERS as a temporary SAS data set.

Imagine the savings when dealing with truly large amounts of data. Suppose, for example, that the raw data reside on four tape volumes. The first DATA step needs another set of tape volumes to hold MEMBERS, and a third set to hold PERM.MEMBERS. The second DATA step stores the view on disk, and needs just a second set of tapes to hold PERM.MEMBERS.

You do not necessarily save CPU time by creating a view. The raw data must be read eventually. If MEMBERS were a view, this program would waste a tremendous amount of CPU time:

```
PROC MEANS DATA=MEMBERS;
CLASS GENDER;
VAR VAR1-VAR20;

PROC FREQ DATA=MEMBERS;
TABLES AGE;

PROC PRINT DATA=MEMBERS;
WHERE AGE > 85;
VAR GENDER VAR18 VAR19;
```

Now each procedure reads from the raw data, a much more expensive task compared to reading from a SAS data set. If the program required just one use of the view MEMBERS, using the view would save on storage space while using only slightly more CPU time.

The final PROC PRINT introduces an interesting feature. Normally, the WHERE statement requires a SAS data set, not raw data. However, even though PROC PRINT reads observations from raw data, the WHERE statement works. Storing a view of raw data enables you to treat raw data as if it were a SAS data set. Consider other features that apply to SAS data sets, such as BY variables and detecting end-of-file when subsetting with a WHERE statement. Storing a view makes these features available when reading from raw data. This example assumes that MEMBERS is a view, based on raw data that are sorted BY LASTNAME:

```
DATA SUBSET;
SET MEMBERS END=NOMORE;
BY LASTNAME;
WHERE AGE > 85;
IF FIRST.LASTNAME THEN TOTAL=0;
TOTAL + VAR15;
IF LAST.LASTNAME;
N + 1;
IF NOMORE THEN CALL SYMPUT('N_NAMES', PUT(N, 5.));
```

This program easily detects the beginning and end of each BY group, as well as end-of-file, despite the fact that the DATA step subsets observations. When subsetting with IF instead of WHERE, those tasks become much more difficult. Even though WHERE cannot be used with raw data, it can be used with the view that extracts from raw data.

CHAPTER 9

Bits and Pieces

Overview

Efficiency encompasses more than minimizing CPU time and storage space. Here are a few remaining topics that have not fit conveniently into earlier chapters.

Gain the Maximum Flexibility

The more choices you have, the more opportunity you have to select a faster programming technique. The more you know, the more choices you have. So spend the time on a regular basis to enhance your programming skills. This can mean reading books, testing programs, and taking courses. However, other methods exist as well. Consider attending users group meetings, not only to learn but to present a paper. When you take the time to write a paper, you sweat the details. It really ingrains your knowledge of the software when you thoroughly research even an introductory level topic and when you look up answers to the questions you anticipate receiving. (Remember, users group meetings welcome introductory as well as advanced level presentations.) On a less formal note, volunteer to help those in your company who know less than you do. Put yourself in a position where you have a stake in being thorough and accurate.

When you begin a project, understand the requirements thoroughly. Talk with the client to determine which features are absolute necessities and which have some flexibility. If you see that changing a specification will markedly improve efficiency, inform the client. Reading the written specifications will not tell you all you need to know!

Compilation Issues

By definition, savings on compilation relate to the length of the program, not to the amount of data. Unless you are running hundreds of lines of code on a regular basis, ignore this topic.

You can precompile an entire DATA step by modifying the RUN statement:

```
DATA NEW;
SET OLD;
CUPS = 2*PINTS + 4*QUARTS;
RUN PGM=COMPILED.CODE;
```

This saves the software the work of interpreting your statements and checking syntax when you run the code. You cannot precompile PROC steps because they are already precompiled modules. However, you can precompile definitions of macros by adding a few statements to the program:

```
LIBNAME ANYNAME 'SAS data library';
OPTIONS MSTORED SASMSTORE=ANYNAME;

%MACRO MYMAC / STORE;
   /* macro definition */
%MEND MYMAC;
```

For more information, refer to the SAS Institute publication *SAS Macro Language: Reference, First Edition*, pages 108-109.

If your program does not use macro language, it will compile faster with the macro processor turned off. When the macro processor is turned on, it examines each word in the program to see whether that word contains macro language. When a word contains text only, the macro processor then passes that word along to be compiled as the next word in the program. When your program uses no macro language, this examination by the macro processor represents an extra step that you can eliminate by specifying the NOMACRO option at SAS invocation.

Finally, embedded comments (using /* and */) compile faster than comment statements (beginning with *). Along similar lines, when a macro executes, macro language comment statements (beginning with %*) compile faster than SAS language comments. If a macro generates a SAS comment, the software must then compile that generated comment.

I/O vs. CPU Time

Most of this book focuses on CPU time, ignoring the fact that I/O is actually a different resource. It is true that many factors that speed up CPU time also improve I/O. Some of those factors mentioned in this book include:

- **Reading data.**
  Read in just the variables needed for the current program. If subsetting observations from raw data, read in just the variables needed to subset; then read the remaining variables for observations that pass the subsetting criteria. Store permanent SAS data sets instead of returning to the raw data every time.

- **Reporting.**
  Use a _NULL_ DATA step when possible.

- **File handling.**
  Use PROC SQL to combine multiple data sets using different keys.

- **Sorting.**
  Sort only when necessary, sorting the required variables only.

- **Summarizing data.**
  Save permanent summaries of SAS data sets, and resummarize those summary data sets as needed.

- **Manipulating data.**
  Use temporary arrays instead of variables to hold a set of numeric constants.

- **Storage space.**
  Shrink the size of a SAS data set (drop variables, reduce variable lengths, compress the data set).

I/O is really a separate resource from CPU time. Even when CPU time is low, I/O can be a bottleneck, or it can be the more expensive resource in a chargeback system. So some mention of tools and techniques related specifically to I/O is warranted.

The WHERE statement looks through a series of observations, without moving the data, until it finds one that meets the WHERE condition(s). Compared to a subsetting IF, the WHERE statement decreases I/O whether it decreases or increases CPU time.

By reusing available disk space, the SAS System can store an individual SAS data set in a series of different physical locations. In that case, moving the fragmented blocks of data will use more I/O. To remedy this situation, use PROC COPY to copy the library to a new location.

System-dependent options, such as BUFNO=, can dramatically affect I/O. For details on this and other options under the MVS operating system, refer to Mike Raithel's book, *Tuning SAS Applications in the MVS Environment*, published by SAS Institute. For operating-system-specific tips, the SAS companion for your operating system may offer additional tips.

Finally, beware of retrieving an entire data set via an index. As noted in Chapter 4, this usually increases I/O and CPU time dramatically.

Plan the Entire Program

Planning helps you determine exactly what work the program must perform. For example, plan which variables and observations will be needed. Then modify your program to read in only the needed data. The clearest, fastest programs perform exactly what is needed.

For an individual DATA step, it's easy to copy from an existing program or even to %INCLUDE a long INPUT statement. But if your DATA step reads in more data than are needed, it becomes expensive in two ways:

- The program takes longer to run.

- The program takes more time to maintain. A programmer (possibly yourself) who works with the program in the future might question why the program reads in certain variables. Sometimes the answer is that the original programmer was too lazy to remove those variables from the program.

Bad planning adds an extra step to this program:

```
DATA MILK;
SET ALLCOWS;
TOTCUPS = 2*PINTS + 4*QUARTS;

PROC MEANS DATA=MILK;
CLASS COW;
VAR TOTCUPS;

DATA LOCATION;
SET ALLCOWS;
IF SITE='HOME' THEN GRASS='GREEN   ';
ELSE                 GRASS='GREENER';

PROC FREQ DATA=LOCATION;
TABLES GRASS;
```

One DATA step could have performed all the data manipulation needed for both PROC MEANS and PROC FREQ:

```
DATA MILK;
SET ALLCOWS;
TOTCUPS = 2*PINTS + 4*QUARTS;
IF SITE='HOME' THEN GRASS='GREEN   ';
ELSE                 GRASS='GREENER';
```

Read and write the data once instead of twice! While this example is simple, the principle applies to complex programs as well. Map out all the reports (or other outputs) that the program must produce. Plan a data manipulation strategy that reads and writes the data a minimum number of times.

Low Maintenance Programming Techniques

Programs take time to write and maintain. Of course, time is money. The previous section mentioned how planning out the program reduces maintenance by eliminating unnecessary steps. Let's examine some additional techniques that reduce maintenance time by making programs easier to understand.

Low Maintenance: Comments

Add comments to document programs. These features indicate a need for more comments:

- The longer you expect a program to remain in use, the more comment statements should be added. One-time programs might need no comments. Programs which will remain in use for extended periods should document with extensive comments.

- The greater the number of people who will run the program, the greater the need for comments (especially when one of the users is a client).

- When the program uses a tricky programming technique or takes an unusual step, add a comment. Even if you are the only person who will ever run the program, these comments will help six months down the road, when you are no longer so familiar with what the program is doing and why.

Begin each program with a lengthy comment, documenting at a minimum the program name, creation date, and purpose. Consider adding the programmer's name, the project, program inputs and outputs, relationship of the program to other programs, and modifications to the program (with their dates). An initial comment might look like this:

```
********************************************************
**                                                    **
**    Program:     C:\DIR1\DIR2\  READDATA.SAS         **
**                                                    **
**    Project:     National User Survey                **
**                                                    **
**    Written by:  Bob Virgile                         **
**                                                    **
**    Date:        March 8, 1999                       **
**                                                    **
**    Purpose:     Read in raw survey data, and        **
**                 store as permanent SAS file         **
**                                                    **
**    Input(s):    C:\SURVEY\RAWDATA.DAT               **
**                                                    **
**    Output(s):   C:\DIR1\DIR2\SASDATA\SURVEY.SD2     **
**                                                    **
********************************************************;
```

Finally, consider adding a member named DOCUMENT (or DOCU-MENT.SAS) in all program libraries. It should briefly summarize the purpose of each program within the library.

Low Maintenance: Programming Style

Many elements of programming style affect readability and, therefore, maintenance time. To begin with, use meaningful names for variables, data sets, filenames, and libnames. Consider which INFILE statement and which PROC MEANS is easier to understand:

```
INFILE RAWDATA1;
INFILE PAYMENTS;

PROC MEANS DATA=IN.FILE;
VAR V1;

PROC MEANS DATA=USA.SALES;
VAR AMOUNT;
```

When you select meaningful names, be consistent. Both DOB and D_BIRTH are reasonable variable names for date of birth. Whichever you select, use that same name in all programs. To compensate for using short, easy to read variable names, label all variables. Many procedures automatically print the variable labels, as well as the names. PROC CONTENTS generates an easy to read description of an entire SAS data set, including the labels of both the variables and the data set.

Specify data set names in all PROC statements. It is much easier to follow the flow of information through a program when all procedures add the DATA= option.

Use spacing and indention to help document your programs. While the following program is relatively simple, the lack of spacing makes it difficult to read:

```
DATA CITIES;
INFILE RAW;
INPUT CITY$15. STATE$2. POP 8. INCOME 6.;
IF POP > 500000;
PROC SORT DATA=CITIES;BY STATE;
PROC UNIVARIATE DATA=CITIES;BY STATE;
VAR INCOME;
```

Spacing could make this program easier to read, while using identical statements:

```
DATA CITIES;
INFILE RAW;
INPUT CITY    $15.
      STATE   $ 2.
      POP       8.
      INCOME    6.;
IF POP > 500000;

PROC SORT DATA=CITIES;
BY STATE;

PROC UNIVARIATE DATA=CITIES;
BY STATE;
VAR INCOME;
```

Some elements of style are really matters of personal taste. Some programmers like to indent every statement within DATA and PROC steps. Some like to indent the word DO. Some like to line up an END statement with the matching DO, and some like to indent the END even further. For these elements of style, don't agonize over which style is better. It is more important that you select one style and stick to it. Sticking with one style makes your programs as a whole easier to understand.

Low Maintenance: Programming Technique

Sometimes slight changes to a program can affect efficiency while still generating the same outcome. The three DATA steps below, for example, each calculate subtotals:

```
PROC SORT DATA=BACTERIA;
BY STRAIN;

DATA SUBTOT1;
SET BACTERIA;
BY STRAIN;
IF FIRST.STRAIN THEN TOTAL=0;
TOTAL + TOXINS;
IF LAST.STRAIN;

DATA SUBTOT2;
SET BACTERIA;
BY STRAIN;
IF FIRST.STRAIN THEN TOTAL=TOXINS;
ELSE TOTAL + TOXINS;
IF LAST.STRAIN;

DATA SUBTOT3;
SET BACTERIA;
BY STRAIN;
TOTAL + TOXINS;
IF LAST.STRAIN;
OUTPUT;
TOTAL=0;
```

Even though one DATA step might run slightly faster than another, it is more important that you select one technique and stick to it. You will find that, as a whole, your programs become easier to work with when they all use similar programming techniques.

Finally, modularize your code to lessen future maintenance. At the extreme, this might mean setting up libraries of macros or source code. But simple steps can also modularize programs, as long as the intention is to make one section of code perform a given task. Consider this section of a DATA step:

```
IF AGE < 21 THEN DO;
   DIFFER  = AGE - 21;
   DRINKER = 'NO ';
   STATUS  = 'MINOR';
END;
ELSE DO;
   DIFFER  = AGE - 21;
   DRINKER = 'YES';
   STATUS  = 'MAJOR';
END;
```

The two statements in bold type are identical. Therefore, one statement should perform that task instead of two. Simply place the statement outside of any DO groups. Let's take one more example from a DATA step:

```
IF HEALTH='OK' THEN PUT
   @1   'Health OK' @12 AGE 3.
   @20 HEIGHT 2.    @25 WEIGHT 3.
   @30 EXERCISE $1. @33 DRINK $1.
   @36 SMOKE $1.;
ELSE PUT
   @1   ILLNESS     @12 AGE 3.
   @20 HEIGHT 2.    @25 WEIGHT 3.
   @30 EXERCISE $1. @33 DRINK $1.
   @36 SMOKE $1.;
```

Once again, the portions in bold type are identical. A low maintenance program would write those variables using a single statement:

```
IF HEALTH='OK' THEN PUT @1 'Health OK' @;
ELSE PUT @1 ILLNESS @;
PUT @12 AGE 3.
    @20 HEIGHT 2.    @25 WEIGHT 3.
    @30 EXERCISE $1.   @33 DRINK $1.
    @36 SMOKE $1.;
```

This change clarifies which portions of the report are identical regardless of the value of HEALTH. In addition, if the format of the report changes, you can update one statement instead of two.

Rather than creating the same format in different programs, save formats permanently. For example, this program saves the REGION format in the catalog PERM.FORMATS by adding the LIBRARY= option:

```
LIBNAME PERM 'path to SAS data library';

PROC FORMAT LIBRARY=PERM;
VALUE REGION   1='North'
               2='South'
               3='East'
               4='West';
```

Dozens of later programs can access all formats in the catalog by setting the FMTSEARCH option:

```
LIBNAME SOMENAME 'path to same SAS data library';
OPTIONS FMTSEARCH=(SOMENAME);
```

Now the later programs don't have to create the format. Equally important, if a format ever changes, you can update the permanent format just once. All the later programs are automatically updated when the format changes.

Choose the Right Tool for the Job

If your program needs to sort, use PROC SORT. If it needs to count, use PROC FREQ. If it needs to compute statistics, use PROC MEANS. Choose the tool that is built for the job, even if you can force the software to complete the task in some other way. Even though you can force PROC SQL to sort a data set, PROC SORT is both faster and clearer:

```
PROC SORT DATA=SALES;
BY STATE;

PROC SQL;
CREATE TABLE SALES AS SELECT * FROM SALES
ORDER BY STATE;
```

You can force PROC FREQ to compute a sum, replacing a PROC MEANS:

```
PROC MEANS DATA=SALES SUM;
VAR AMOUNT;

DATA TEST;
SET SALES;
RETAIN J 1;

PROC FREQ DATA=TEST;
TABLES J;
WEIGHT AMOUNT;
```

Because PROC MEANS is built to calculate statistics, it will both run faster (even without the extra DATA step to add the variable J) and be easier to read. You can also force PROC MEANS to count, replacing PROC FREQ:

```
PROC FREQ DATA=SALES;
TABLES STATE;

DATA TEST;
SET SALES;
RETAIN J 1;

PROC MEANS DATA=SALES N;
CLASS STATE;
VAR J;
```

PROC FREQ works better here because it is built to count. Even if PROC MEANS runs faster (and it occasionally has under some releases of the software), it becomes more confusing to read. The only good time to get counts out of PROC MEANS is when you need PROC MEANS to calculate other statistics at the same time. In that case, remember that the N statistic is NOT the number of observations but rather is the number of observations where the analysis variable (J) has a nonmissing value. The automatic statistic _FREQ_ is the number of observations for each STATE.

CHAPTER 10

Test Programs

Designing Test Programs

Test programs should isolate the alternative statements, eliminating other work that the programs perform. Suppose, for example, we wanted to compare these two statements:

```
INPUT @1 ZIPCODE 5.;
INPUT @1 ZIPCODE $CHAR5.;
```

The following alternatives perform too much work:

```
DATA TEST1;
INFILE RAWDATA;
INPUT @1 ZIPCODE 5.;
```

```
DATA TEST2;
INFILE RAWDATA;
INPUT @1 ZIPCODE $CHAR5.;
```

In addition to reading in ZIPCODE, these programs perform other work that obscures the relative speed of the INPUT statements. First of all, both programs output observations to a new data set. Instead, the programs should use a DATA _NULL_; statement to prevent outputting any observations. Secondly, both programs bring in blocks of raw data. To eliminate that work, a DATA step could read a single line of data many times. A better test would be:

```
DATA _NULL_;
DO I=1 TO 500000;
    INPUT @1 ZIPCODE 5. @;
END;
CARDS;
01801
;

DATA _NULL_;
DO I=1 TO 500000;
    INPUT @1 ZIPCODE $CHAR5. @;
END;
CARDS;
01801
;
```

Even with these improvements, the CPU times are misleading. For example, suppose the first DATA step took twice the CPU time of the second. It would be misleading to say that the $CHAR5. informat reads data twice as fast as the 5. informat. The CPU time for both DATA steps includes the overhead of running through a DO loop and an INPUT statement 500,000 times. To obtain true percentage savings, that overhead should be subtracted out. One final DATA step supplies this baseline measurement:

```
DATA _NULL_;
DO I=1 TO 500000;
    INPUT @;
END;
STOP;
CARDS;
01801
;
```

A complete set of test programs must run such a baseline test, measuring and removing all overhead from the other programs. The section "Calculating Relative Performance" explains this in more depth.

Benefits of Testing with DO Loops

DO loops produce consistent, reliable results. On the other hand, programs that read multiple lines of data can vary unpredictably from one test to the next. (I occasionally use the term "random fluctuations" to refer to these unpredictable variations in CPU time.) When you test the same program three times, it is comforting to see it use the same CPU time on each run.

By testing in a DO loop, you also increase flexibility. If you are testing on a mainframe and using the latest release of the software, you can easily increase the number of iterations to 5,000,000. On the other hand, if you are chugging along using an older PC with an earlier release of the software, you can just as easily decrease the number of iterations to 50,000. You don't have to locate a data set of just the right size. Instead, just change the upper bound in the DO statement.

Benefits of Testing on Multiple Observations

Either method (DO loops or reading in data) can subtract out program overhead to compare alternative programming techniques. However, in real life, the vast majority of DATA steps read in data. Therefore, tests that read in data give you a better idea of a statement's impact on the entire program. The program's overhead becomes useful information, by placing the savings within the context of a larger, more realistic program.

Some tests require reading many lines of raw data because the savings center around work that takes place between observations.

For example, consider these two DATA steps:

```
DATA TEST1;
INFILE RAWDATA;
INPUT VAR1 3.;

DATA TEST2;
INFILE RAWDATA;
INPUT VAR1 3.;
RETAIN VAR1;
```

While the output data sets are identical, the first DATA step uses a small amount of CPU time to reset VAR1 to missing at the beginning of each observation. To measure this CPU time, a test program cannot read a single line of data multiple times in a DO loop. Instead, DATA steps must move on from one observation to the next to allow the software to reset VAR1 to missing. Even the previous test, comparing the 5. and $CHAR5. informats, could have read many lines of raw data. For example, the first of these three DATA steps provides a baseline measurement:

```
DATA _NULL_;
INFILE RAWDATA;
INPUT;

DATA _NULL_;
INFILE RAWDATA;
INPUT ZIPCODE 5.;

DATA _NULL_;
INFILE RAWDATA;
INPUT ZIPCODE $CHAR5.;
```

Clearly, by reading multiple lines of data, you can compare the 5. and the $CHAR5. informats. Because some tests must read multiple raw data lines, while other tests could have used DO loops, many of the test runs read that data whether or not they could have used DO loops instead.

Calculating Relative Performance

In the DATA steps above, the first DATA step measures the overhead incurred by reading through all the data lines. A test that compares the 5. informat to the $CHAR5. informat should report the CPU time for all three DATA steps. In that case, the true relative efficiency would be:

$$\frac{(\text{CPU time for 2nd DATA step}) - (\text{CPU time for 1st DATA step})}{(\text{CPU time for 3rd DATA step}) - (\text{CPU time for 1st DATA step})}$$

This ratio compares the marginal effect of adding a variable to the INPUT statement, using different informats. The ratio does not measure the effect on the entire program, the entire DATA step, or an entire INPUT statement that reads many variables. To assess performance in these other contexts, you would need to compare the total CPU time for all three DATA steps.

Creating Test Data

You can easily create a test data set with specific dimensions. For example, this DATA step creates a test data set holding 30 variables and 500,000 observations:

```
DATA _030VARS;
DO VAR1=1 TO 500000;
   OUTPUT;
END;
RETAIN VAR2-VAR15 0 VAR16-VAR30 'Richard';
```

If you need more variation in the test data, you can modify the RETAIN statement and/or add statements inside the DO loop, such as:

```
IF VAR1=100001 THEN VAR30='Michael';
```

In the test programs that follow, the data set _030VARS always contains 30 variables and 100,000 observations, while _100VARS always contains 100 variables and 100,000 observations.

The Operating Systems

All test programs ran on four operating systems: MVS, Windows NT, UNIX, and OpenVMS. While these operating systems are reasonably comprehensive, you can always download and run the same test programs on your hardware. Here are some of the details regarding the test equipment.

MVS, SAS 6.09.0455P042397

The mainframes consisted of a set of three IBM Model 9672 machines.

Windows NT, SAS 6.12 TS020

The Windows NT Server was a Hewlett-Packard NetserverLX 4-way (200 mhz processors) with 1 gigabyte of RAM and 80 gigabytes of disk space, running NT 4.0 (service pack level 3).

UNIX, SAS 6.12 TS020

The Sun Server was an Enterprise 4000 8-way (167 mhz processors) with 1 gigabyte of RAM and 120 gigabytes of disk space, running Solaris 2.5.

OpenVMS, SAS 6.12 TS020

The Alpha server was an AlphaServer 4100 5/466 2-way, with 1 gigabyte of RAM and 20 gigabytes of disk space, running Open VMS 7.1.

On Windows NT, the SAS System reports total elapsed time, not CPU time. In many cases, the results would be comparable. However, because total elapsed time includes I/O time, processes that use large amounts of I/O will appear to take relatively longer under Windows NT. For example, in one sample test, PROC SORT completed faster than PROC MEANS on the other three operating systems, but it took much longer than PROC MEANS on Windows NT. The particular PC running Windows NT had one channel available for both input and output, compounding this problem.

Reporting the Test Results

Test results appear as percentages, with a baseline technique using 100% of the CPU time on each operating system. Nearly all tests are based on the total CPU time needed to run a test program three times. Rarely, the CPU time for one test run of three differed drastically from the CPU time on the other two runs. Whether the reasons for the difference could be determined or not, whenever the CPU time for a run differed by more than 40% from the average of the other two runs, the drastically different result was thrown out and replaced by the average time of the other two tests.

Finally, each test also lists

- relevant conditions that exist in the data,

- interesting (to me) results, based on the test numbers.

- any patterns or variations in the results that might have been obscured by summarizing the data and presenting percentages.

Results for Chapter 2: Reading Data

Test 2A: List vs. Formatted vs. Column INPUT

In the raw data, column 6 is always blank.

```
A: DATA _NULL_;
   INFILE RAWDATA;
   INPUT;

B: DATA _NULL_;
   INFILE RAWDATA;
   INPUT @4 AGE;

C: DATA _NULL_;
   INFILE RAWDATA;
   INPUT @4 AGE 2.;
```

D: `DATA _NULL_;`
   `INFILE RAWDATA;`
   `INPUT AGE 4-5;`

	Operating System			
Test	MVS	Win NT	UNIX	OpenVMS
A: Null INPUT	100%	100%	100%	100%
B: List INPUT	227%	162%	136%	130%
C: Formatted INPUT	172%	138%	130%	119%
D: Column INPUT	173%	137%	125%	116%

Notes:

- As expected, list input is always slowest.

- Column and formatted input use equal amounts of CPU time.

- The cost of adding variables to the INPUT statement is relatively lower under UNIX and VMS. Of course, you can interpret this result in two ways: either a null INPUT statement is slow on these platforms or the process of reading variables is fast. In practical terms, these numbers mean that improvements to the INPUT statement will have less of an impact on the entire DATA step for these two operating systems.

- As your programs start reading in more variables, the cost of the null INPUT statement becomes less and less significant. While the next test continues to report on the null INPUT statement as a baseline measurement, later tests will not.

Test 2B: 5. vs. $5. vs. $CHAR5. Informats

The test data do not contain decimal points. While this might make a difference, in real life it should not be a consideration. There are very few numerals that contain decimal points where it makes any kind of sense to read the data values as character.

```
A: DATA _NULL_;
   INFILE RAWDATA;
   INPUT;

B: DATA _NULL_;
   INFILE RAWDATA;
   INPUT ZIPCODE 5.;

C: DATA _NULL_;
   INFILE RAWDATA;
   INPUT ZIPCODE $5.;

D: DATA _NULL_;
   INFILE RAWDATA;
   INPUT ZIPCODE $CHAR5.;
```

	Operating System			
Test	MVS	Win NT	UNIX	OpenVMS
A: Null INPUT	100%	100%	100%	100%
B: 5. Informat	173%	147%	126%	119%
C: $5. Informat	182%	149%	125%	130%
D: $CHAR5. Informat	134%	118%	111%	121%

Notes:

- Reading variables as character ($5.) vs. numeric (5.) makes little difference. In fact, most operating systems read data faster as numeric!

- The $CHAR5. informat reads data twice as fast as the $5. informat. For example, under MVS the added CPU time to read a variable with the $5. informat is 182% - 100% = 82%. The added CPU time for the $CHAR5. informat is 134% - 100% = 34%.

- Quite surprisingly, Open VMS reads data fastest as numeric.

- By using _NULL_ DATA steps, the programs output nothing. Therefore, these tests measure differences in CPU time needed to read, but not to output, a numeric vs. a character variable. If you leave your numeric variables with their default length of 8 bytes, expect that it will take longer to output 8 bytes than to output 5.

Test 2C: Remove Starting Columns When Reading Consecutive Fields

The test INPUT statements contained exactly the same pattern of dollar signs as appear below. If the order appears a little odd, that is merely a reflection of the values in the raw data.

```
A: INPUT @1 V1 1. @2 V2 1. @3 V3 1. @4 V4 1. @5 V5 1. @6 V6 $1.
         @7 V7 $1. @8 V8 $1. @9 V9 $1. @10 V10 $1. @11 V11 $1.
         @12 V12 $1. @13 V13 1. @14 V14 1. @15 V15 1.
         @16 V16 $1.  @17 V17 $1. @18 V18 $1. @19 V19 $1.
         @20 V20 $1.;

B: INPUT V1 1. V2 1. V3 1. V4 1. V5 1. V6 $1. V7 $1. V8 $1.
         V9 $1. V10 $1. V11 $1. V12 $1. V13 1. V14 1. V15 1.
         V16 $1.  V17 $1. V18 $1. V19 $1. V20 $1.;
```

	Operating System			
Test	MVS	Win NT	UNIX	OpenVMS
A: Starting columns	100%	100%	100%	100%
B: No starting columns	85%	78%	83%	91%

Notes:

- At last, consistent results across operating systems! The percentage savings are high because the INPUT statements read just one byte per variable. In practice, expect the percentage savings to be smaller but consistent.

- These numbers represent net CPU time. For example, the baseline of 100% in method A is really the CPU time to execute the DATA step including the INPUT statement, minus the CPU time to execute a baseline DATA step with a null INPUT statement. All future tests use this same format.

Test 2D: Conversions with an Informat vs. the INPUT Function

A: `INPUT @14 SASDATE YYMMDD6.;`

B: `INPUT @14 CHARDATE $CHAR6.;`
   `SASDATE = INPUT(CHARDATE, YYMMDD6.);`

	Operating System			
Test	MVS	Win NT	UNIX	OpenVMS
A: Informat	100%	100%	100%	100%
B: INPUT function	106%	115%	107%	126%

Notes:

- On all operating systems, one statement runs faster than two.

- Even when the numbers are close, such as 106% and 107%, the results are consistent. For any operating system, each test run for method A completed faster than each test run for method B.

Test 2E: User-Defined Informat vs. IF/THEN

Method B first created an informat using this program:

```
PROC FORMAT;
INVALUE $GENDER 'MALE'=1 'FEMALE'=2;
```

The table below disregards the CPU time needed to create this informat.

A: `INPUT @7 TYPE $GENDER6.;`

B:
```
INPUT @7 GENDER $CHAR6.;
IF GENDER='MALE' THEN TYPE=1;
ELSE TYPE=2;
```

	Operating System			
Test	MVS	Win NT	UNIX	OpenVMS
A: Informat	100%	100%	100%	100%
B: IF THEN statements	23%	20%	39%	65%

Notes:

- IF/THEN statements apparently run incredibly quickly! The percentage savings vary, but IF/THEN always runs considerably faster than the informat.

- Perhaps the final statement should be more explicit to make the two tests identical:

```
ELSE IF GENDER='FEMALE' THEN TYPE=2;
```

- Chapter 7 examines a related test (7M), where IF/THEN statements outperform a format for recoding variables. In that test, the format contains 26 categories, instead of the two in this informat.

Test 2F.1: Subsetting Variables in PROC SORT

All PROC SORTs use the OUT= option to avoid replacing the original data set with a sorted version. As a result, the same program can run three times on the original data.

```
A: PROC SORT DATA=_030VARS OUT=TEMP;
   BY VAR1;

B: PROC SORT DATA=_030VARS OUT=TEMP (KEEP=VAR1-VAR3);
   BY VAR1;

C: PROC SORT DATA=_030VARS (KEEP=VAR1-VAR3) OUT=TEMP;
   BY VAR1;
```

	Operating System			
Test	MVS	Win NT	UNIX	OpenVMS
A: All variables	100%	100%	100%	100%
B: KEEP= on output	89%	55%	69%	89%
C: KEEP= on input	89%	58%	68%	89%

Notes:

- Shifting the (KEEP=) data set option from the output to the input data set has no effect! In both cases (tests B and C), Version 6 sorts all variables, subsetting them when writing to TEMP. Version 7 supports subsetting at input or output.

Test 2F.2: Subsetting Variables in PROC MEANS

The data set TEMP contains three variables, including VAR1 and VAR3.

```
A: PROC MEANS DATA=_030VARS;
   VAR VAR1 VAR3;
```

```
B: PROC MEANS DATA=_030VARS (KEEP=VAR1-VAR3);
   VAR VAR1 VAR3;
```

```
C: PROC MEANS DATA=TEMP;
   VAR VAR1 VAR3;
```

	Operating System			
Test	MVS	Win NT	UNIX	OpenVMS
A: PROC MEANS, analyze wide data set	100%	100%	100%	100%
B: PROC MEANS, KEEP=, wide data set	100%	108%	105%	95%
C: PROC MEANS, analyze narrow data set	88%	82%	85%	75%

Notes:

- The KEEP= data set option has no effect. However, performing the same analysis on a "skinnier" data set consistently reduces the CPU time.

Test 2G: Subsetting Observations in a DATA Step

In a data set with 100,000 observations, VAR1 takes on integer values from 1 through 100,000. This first set of tests, 2G.1 through 2G.4, subsets from a data set containing 30 variables. The second set of tests, 2G.5 through 2G.8, subsets from a data set containing 100 variables.

The DELETE statement uses a different cutoff to avoid a compound operator. Checking for VAR1 < 80001 clearly makes one comparison. Checking for VAR1 <= 80000 might conceivably be making two comparisons.

Test 2G.1: Taking a 20% Subset, with 30 Variables

```
A: DATA _NULL_;
     SET _030VARS;
     IF VAR1 > 80000;

B: DATA _NULL_;
     SET _030VARS;
     IF VAR1 < 80001 THEN DELETE;

C: DATA _NULL_
     SET _030VARS;
     WHERE VAR1 > 80000;
```

	Operating System			
Test	MVS	Win NT	UNIX	OpenVMS
A: Subsetting IF	100%	100%	100%	100%
B: DELETE statement	101%	101%	99%	102%
C: WHERE statement	95%	79%	67%	78%

Test 2G.2: Taking a 30% Subset, with 30 Variables

A: `IF VAR1 > 70000;`

B: `IF VAR1 < 70001 THEN DELETE;`

C: `WHERE VAR1 > 70000;`

Test		Operating System		
	MVS	Win NT	UNIX	OpenVMS
A: Subsetting IF	100%	100%	100%	100%
B: DELETE statement	99%	98%	100%	90%
C: WHERE statement	101%	89%	74%	80%

Test 2G.3: Taking a 50% Subset, with 30 Variables

A: `IF VAR1 > 50000;`

B: `IF VAR1 < 50001 THEN DELETE;`

C: `WHERE VAR1 > 50000;`

Test		Operating System		
	MVS	Win NT	UNIX	OpenVMS
A: Subsetting IF	100%	100%	100%	100%
B: DELETE statement	99%	97%	99%	115%
C: WHERE statement	117%	104%	89%	94%

Test 2G.4: Taking an 80% Subset, with 30 Variables

A: `IF VAR1 > 20000;`

B: `IF VAR1 < 20001 THEN DELETE;`

C: `WHERE VAR1 > 20000;`

		Operating System		
Test	MVS	Win NT	UNIX	OpenVMS
A: Subsetting IF	100%	100%	100%	100%
B: DELETE statement	99%	100%	100%	84%
C: WHERE statement	136%	115%	112%	118%

Notes:

- As expected, the subsetting IF and DELETE statements produce similar results. There is no explanation for the fluctuation in results on OpenVMS.

- As expected, WHERE becomes less efficient as the selected sample size grows.

Test 2G.5: Taking a 20% Subset, with 100 Variables

A: `IF VAR1 > 80000;`

B: `IF VAR1 < 80001 THEN DELETE;`

C: `WHERE VAR1 > 80000;`

	Operating System			
Test	MVS	Win NT	UNIX	OpenVMS
A: Subsetting IF	100%	100%	100%	100%
B: DELETE statement	99%	94%	99%	101%
C: WHERE statement	74%	71%	62%	86%

Test 2G.6: Taking a 30% Subset, with 100 Variables

A: `IF VAR1 > 70000;`

B: `IF VAR1 < 70001 THEN DELETE;`

C: `WHERE VAR1 > 70000;`

	Operating System			
Test	MVS	Win NT	UNIX	OpenVMS
A: Subsetting IF	100%	100%	100%	100%
B: DELETE statement	100%	93%	97%	93%
C: WHERE statement	83%	76%	66%	79%

Test 2G.7: Taking a 50% Subset, with 100 Variables

A: `IF VAR1 > 50000;`

B: `IF VAR1 < 50001 THEN DELETE;`

C: `WHERE VAR1 > 50000;`

	Operating System			
Test	MVS	Win NT	UNIX	OpenVMS
A: Subsetting IF	100%	100%	100%	100%
B: DELETE statement	100%	88%	97%	104%
C: WHERE statement	96%	80%	78%	91%

Test 2G.8: Taking an 80% Subset, with 100 Variables

A: `IF VAR1 > 20000;`

B: `IF VAR1 < 20001 THEN DELETE;`

C: `WHERE VAR1 > 20000;`

	Operating System			
Test	MVS	Win NT	UNIX	OpenVMS
A: Subsetting IF	100%	100%	100%	100%
B: DELETE statement	100%	92%	97%	102%
C: WHERE statement	115%	96%	97%	98%

Notes:

- With the larger number of variables, WHERE remains competitive even for large sample sizes.

- Using a _NULL_ DATA step produces a curious result for test methods A and B. Each test (2G.5 through 2G.8) reads in all observations and outputs nothing. Therefore, the CPU time for all four tests remains the same, regardless of the cutoff point for deleting observations.

Test 2H.1: Beating the DATA Step Loop, Reading a SAS File

```
A: DATA _NULL_;
   SET _100VARS;
```

```
B: DATA _NULL_;
   DO UNTIL (NOMORE);
      SET _100VARS END=NOMORE;
   END;
```

	Operating System			
Test	MVS	Win NT	UNIX	OpenVMS
A: Baseline DATA step	100%	100%	100%	100%
B: DO Loop	97%	99%	99%	98%

Notes:

- The savings are minimal, perhaps nonexistent.

- Without a _NULL_ DATA step, the second program would have to add an OUTPUT statement within the DO loop to generate identical results.

Test 2H.2: Beating the DATA Step Loop, Reading Raw Data

```
A: DATA _NULL_;
   INFILE RAWDATA;
   INPUT (VAR1-VAR20) ($1.);
```

```
B: DATA _NULL_;
   INFILE RAWDATA;
   INPUT (VAR1-VAR20) ($1.);
   RETAIN VAR1-VAR20;
```

```
C: DATA _NULL_;
   INFILE RAWDATA END=NOMORE;
   DO UNTIL (NOMORE);
      INPUT (VAR1-VAR20) ($1.);
   END;
```

Test	Operating System			
	MVS	Win NT	UNIX	OpenVMS
A: Baseline DATA step	100%	100%	100%	100%
B: RETAIN variables	99%	98%	101%	98%
C: DO loop	107%	101%	103%	98%

Notes:

- The savings from program B, retaining additional variables, is minimal and therefore difficult to measure. Random variation from one test run to the next actually results in slightly higher CPU time on UNIX, even though we "know" the program performs less work when using method B.

- The added overhead of a DO loop wipes out any savings from method C.

Test Results for Chapter 3: Reporting

Test 3A: Controlling Formats when Printing

```
A: PROC PRINT DATA=_100VARS (OBS=5000);
   VAR VAR:;
```

```
B: PROC PRINT DATA=_100VARS (OBS=5000);
   VAR VAR:;
   FORMAT _NUMERIC_ 2. _CHARACTER_ $8. VAR1 7.;
```

	Operating System			
Test	MVS	Win NT	UNIX	OpenVMS
A: No FORMAT statement	100%	100%	100%	100%
B: Add FORMAT statement	26%	38%	31%	27%

Notes:

- All operating systems reveal substantial savings.

Test 3B: Using a _NULL_ Data Step

```
A: DATA TEMP;
   SET _100VARS;

B: DATA _NULL_;
   SET _100VARS;
```

	Operating System			
Test	MVS	Win NT	UNIX	OpenVMS
A: Baseline DATA step	100%	100%	100%	100%
B: DATA _NULL_	52%	4%	31%	53%

Notes:

- MVS and VMS generate the expected savings. Reading an observation should take about the same amount of time as writing an observation. Therefore, a _NULL_ DATA step should cut out nearly half the work of the DATA step.

- The results under Windows NT are shocking. The test was run three times, with the same result all three times. Remember that Windows NT reports total elapsed time and that the PC had one channel available for both input and output. Even so, these numbers are extreme.

Test 3C.1: Writing _INFILE_ vs. One Variable

The incoming raw data lines are 20 characters long, so STRING and _INFILE_ are identical. The PUT statement uses a trailing @ to limit the size of the report while still writing out each line of data. No baseline measurements were run (for example, a similar DATA step without a PUT statement).

```
A: DATA _NULL_;
   INFILE RAWDATA;
   FILE PRINT NOTITLES;
   INPUT STRING $CHAR20.;
   PUT @1 STRING $CHAR20. @;

B: DATA _NULL_;
   INFILE RAWDATA;
   FILE PRINT NOTITLES;
   INPUT STRING $CHAR20.;
   PUT @1 _INFILE_ @;
```

	Operating System			
Test	MVS	Win NT	UNIX	OpenVMS
A: Write 1 variable	100%	100%	100%	100%
B: Write _INFILE_	107%	120%	100%	115%

Test 3C.2: Writing _INFILE_ vs. Four Variables

```
A: DATA _NULL_;
   INFILE RAWDATA;
   FILE PRINT NOTITLES;
```

```
INPUT (S1-S4) ($CHAR5.);
PUT @1 S1 $CHAR5. S2 $CHAR5. S3 $CHAR5. S4 $CHAR5. @;
```

B:
```
DATA _NULL_;
INFILE RAWDATA;
FILE PRINT NOTITLES;
INPUT (S1-S4) ($CHAR5.);
PUT @1 _INFILE_ @;
```

	Operating System			
Test	MVS	Win NT	UNIX	OpenVMS
A: Write 4 variables	100%	100%	100%	100%
B: Write _INFILE_	90%	112%	90%	106%

Notes:

- The software can write a variable faster than it can write _INFILE_. However, as the number of variables increases, writing _INFILE_ becomes more efficient. The cutoff point varies by operating system.

Test 3D: Summarize Before PROC TABULATE

When one program works from the original data and the other works from a summary, the statements within PROC TABULATE must change to force both programs to generate the same report.

A:
```
PROC TABULATE DATA=_100VARS;
CLASS VAR100;
VAR VAR1;
TABLES VAR100, VAR1*(MEAN MIN MAX) / RTS=15;
```

B:
```
PROC MEANS DATA=_100VARS NOPRINT NWAY;
CLASS VAR100;
VAR VAR1;
OUTPUT OUT=TEMP MEAN=MEAN_ MIN=MIN_ MAX=MAX_;
```

```
PROC TABULATE DATA=TEMP;
CLASS VAR100;
VAR MEAN_ MIN_ MAX_;
TABLES VAR100, SUM='VAR1'*(MEAN_ MIN_ MAX_) / RTS=15;
```

	Operating System			
Test	MVS	Win NT	UNIX	OpenVMS
A: TABULATE only	100%	100%	100%	100%
B: MEANS then TABULATE	108%	129%	110%	150%

Notes:

- In Version 5 of the software, this test used to produce opposite results. One lesson here is to retest efficiency as new releases of the software emerge.

Test Results for Chapter 4: File Handling

Test 4A: *Sorting by Unnecessary Variables*

In this data, VAR3 exhibited very little variation. Therefore, both tests performed about the same amount of sorting. However, the second test had to verify that the data were in the proper order.

Once again, PROC SORT uses the OUT= option to leave the original data set intact. In that way, each test can run several times.

A:
```
PROC SORT DATA=_030VARS OUT=TEMP;
BY VAR1;
```

B:
```
PROC SORT DATA=_030VARS OUT=TEMP;
BY VAR1 VAR3;
```

	Operating System			
Test	MVS	Win NT	UNIX	OpenVMS
A: SORT by 1 variable	100%	100%	100%	100%
B: SORT by 2 variables	103%	97%	94%	95%

Notes:

- Obviously, the first program always performs less work than the second. The only conclusion to draw here is that this chapter always reports the actual results, never doctoring the outcome.

Test 4B.1: PROC SQL vs. a SET Statement, All Variables and Observations

A:
```
PROC SQL;
   CREATE TABLE TEMP AS SELECT * FROM _100VARS;
```

B:
```
DATA TEMP;
   SET _100VARS;
```

	Operating System			
Test	MVS	Win NT	UNIX	OpenVMS
A: PROC SQL	100%	100%	100%	100%
B: DATA step	78%	101%	99%	78%

Test 4B.2: PROC SQL vs. a SET Statement, Subset the Variables

A:
```
PROC SQL;
   CREATE TABLE TEMP AS
   SELECT VAR1, VAR2A, VAR3A FROM _100VARS;
```

```
B:  DATA TEMP;
    SET _100VARS (KEEP=VAR1 VAR2A VAR3A);
```

	Operating System			
Test	MVS	Win NT	UNIX	OpenVMS
A: PROC SQL	100%	100%	100%	100%
B: DATA step	68%	91%	85%	77%

Test 4B.3: PROC SQL vs. a SET Statement, Subset the Observations

Just for the record, zero observations meet the WHERE condition.

```
A:  PROC SQL;
    CREATE TABLE TEMP AS SELECT * FROM _100VARS
    WHERE VAR100='NewValue';
```

```
B:  DATA TEMP;
    SET _100VARS;
    WHERE VAR100='NewValue';
```

	Operating System			
Test	MVS	Win NT	UNIX	OpenVMS
A: PROC SQL	100%	100%	100%	100%
B: DATA step	101%	101%	97%	104%

Notes:

- The DATA step meets or beats PROC SQL for common functions.

Test 4B.4: PROC SQL vs. a MERGE Statement

A: ```
PROC SQL;
CREATE TABLE TEMP AS SELECT * FROM
_030VARS FULL JOIN _100VARS ON _030VARS.VAR1=_100VARS.VAR1;
```

B: ```
PROC SORT DATA=_030VARS;
BY VAR1;

PROC SORT DATA=100VARS;
BY VAR1
```

C: Same as program A, run after sorting.

D: ```
DATA TEMP;
MERGE _030VARS _100VARS;
BY VAR1;
```

|  | Operating System | | | |
| --- | --- | --- | --- | --- |
| Test | MVS | Win NT | UNIX | OpenVMS |
| A: SQL, unsorted data | 100% | 100% | 100% | 100% |
| B: SORT data sets | 23% | 69% | 37% | 38% |
| C: SQL, sorted data | 49% | 37% | 51% | 45% |
| D: MERGE, sorted data | 21% | 36% | 40% | 26% |

Notes:

- Once again, the DATA step always meets or beats PROC SQL.

- The CPU time to sort is high under Windows NT. Chapter 5 addresses techniques to reduce the time needed to sort data.
```

- PROC SQL joins data sets twice as fast using sorted (vs. unsorted) data. It pays to sort these data sets to allow PROC SQL to join them faster.

- Similar tests were run for a LEFT JOIN and RIGHT JOIN. Because all numbers were within 1 or 2 percent of the numbers on this table, this chapter omits the results.

Tests Results for Chapter 5: Sorting Data

As usual, the OUT= option in PROC SORT allows the same code to run several times on the same data set. Limited test results (not reported here) found that the OUT= option adds slightly to the CPU time needed for PROC SORT.

Test 5A: *Use the CLASS Statement to Avoid Sorting*

```
A:  PROC MEANS DATA=_100VARS NWAY;
    CLASS VAR100;
    VAR _NUMERIC_;

B:  PROC SORT DATA=_100VARS OUT=TEMP;
    BY VAR1;

C:  PROC MEANS DATA=TEMP;
    BY VAR100;
    VAR _NUMERIC_;
```

	Operating System			
Test	MVS	Win NT	UNIX	OpenVMS
A: MEANS, with CLASS	100%	100%	100%	100%
B: PROC SORT	27%	898%	53%	94%
C: MEANS, with BY	97%	91%	100%	80%

Notes:

- Both statements, BY and CLASS, generate similar results here because VAR100 takes on only a few values. As the number of values in the CLASS variables rises, the spread widens between BY and CLASS.

- Test 6B conducts some similar tests on a data set containing only two variables. The percentages change as the data set dimensions change, but the bottom line remains the same: use BY for sorted data and CLASS for unsorted data.

Tests 5B and 5C: The SORTSIZE= and NOEQUALS Options

```
A: PROC SORT DATA=_100VARS OUT=TEMP;
   BY VAR100;

B: PROC SORT DATA=_100VARS OUT=TEMP SORTSIZE=MAX;
   BY VAR100;

C: PROC SORT DATA=_100VARS OUT=TEMP NOEQUALS;
   BY VAR100;
```

	Operating System			
Test	MVS	Win NT	UNIX	OpenVMS
A: Baseline PROC SORT	100%	100%	100%	100%
B: SORTSIZE=MAX	101%	57%	100%	102%
C: NOEQUALS	101%	102%	91%	91%

Notes:

- When the SORTSIZE= option makes a difference, it can make a huge difference!

- When the NOEQUALS option makes a difference, it makes a small but consistent difference. Under UNIX and OpenVMS, each test of method C completed faster than each test of method A.

Test 5D: Controlling the Sorting Routine

```
A: PROC SORT DATA=_100VARS OUT=TEMP;
   BY VAR100;
   * uses the default value of the sortpgm option;

B: OPTIONS SORTPGM=SAS;
   PROC SORT DATA=_100VARS OUT=TEMP;
   BY VAR1;
   RUN;
   OPTIONS SORTPGM=BEST;
```

	Operating System			
Test	MVS	Win NT	UNIX	OpenVMS
A: Default SORTPGM	100%	100%	100%	100%
B: SORTPGM=SAS	147%	101%	100%	65%

Notes:

- The tests reveal a surprising feature of OpenVMS. The SAS sorting routine runs faster than the system sorting routine, for a data set with 100,000 observations.

Test 5E: A TAGSORT Workaround

```
A: PROC SORT DATA=_100VARS OUT=TEMP TAGSORT;
   BY VAR100;

B: DATA JUSTKEYS;
   SET _100VARS (KEEP=VAR100);
   OBSNO=_N_;
```

```
PROC SORT DATA=JUSTKEYS;
BY VAR100;

DATA TEMP;
SET JUSTKEYS (KEEP=OBSNO);
SET _100VARS POINT=OBSNO;
```

	Operating System			
Test	MVS	Win NT	UNIX	OpenVMS
A: TAGSORT	100%	100%	100%	100%
B: TAGSORT workaround	36%	109%	116%	116%

Notes:

- When the workaround runs faster, it runs much faster!

- The last DATA step consumed the most CPU time in the workaround, regardless of the operating system.

- Consider modifying this test on your own. For example, incorporate other features from this chapter (such as the SORTSIZE= option) when sorting.

Tests Results for Chapter 6: Summarizing Data

Test 6A: PROC MEANS vs. PROC SUMMARY

A:
```
PROC MEANS DATA=_100VARS;
VAR _NUMERIC_;
```

B:
```
PROC SUMMARY DATA=_100VARS PRINT;
VAR _NUMERIC_;
```

	Operating System			
Test	MVS	Win NT	UNIX	OpenVMS
A: PROC MEANS	100%	100%	100%	100%
B: PROC SUMMARY	100%	99%	100%	100%

Notes:

- With such consistent results, the remaining tests in this chapter run PROC MEANS but not PROC SUMMARY.

Test 6B.1: PROC MEANS vs. PROC SQL vs. the DATA Step, Summarize Entire Data Set

```
A: PROC MEANS DATA=_030VARS SUM N;
   VAR VAR1;

B: PROC SQL;
   SELECT SUM(VAR1) AS SUM, N(VAR1) AS N FROM _030VARS;

C: DATA _NULL_;
   SET _030VARS (KEEP=VAR1) END=EOF;
   SUM + VAR1;
   IF VAR1 > .Z THEN N + 1;
   IF EOF;
   FILE PRINT;
   PUT SUM= N=;
```

	Operating System			
Test	MVS	Win NT	UNIX	OpenVMS
A: PROC MEANS	100%	100%	100%	100%
B: PROC SQL	178%	181%	95%	53%
C: DATA step	89%	151%	59%	38%

Notes:

- On each operating system, the slowest method takes about twice as long as the fastest. While the DATA step usually runs fastest, on Windows NT PROC MEANS runs significantly faster.

- PROC SQL may be competitive, but it is never faster than the DATA step.

- The SUM in the DATA step report is zero whenever the analysis variable is missing on every observation. An IF/THEN statement could easily correct that problem.

Test 6B.2: PROC MEANS vs. PROC SQL vs. the DATA Step, Summarize Subgroups

Tests A and B below run on unsorted data. Tests D and F require sorted data. The fastest technique may depend on whether or not the data are already sorted.

The data set TEMP contains just two variables: VAR1 and VAR100.

Notice that test B runs on unsorted data, while test E runs the same code on sorted data.

A:
```
PROC MEANS DATA=TEMP SUM N;
VAR VAR1;
CLASS VAR100;
```

B:
```
PROC SQL;
SELECT SUM(VAR1) AS SUM, N(VAR1) AS N FROM TEMP
GROUP BY VAR100;
```

C:
```
PROC SORT DATA=TEMP OUT=TEMP2;
BY VAR100;
```

D:
```
DATA _NULL_;
SET TEMP2 END=EOF;
BY VAR100;
SUM + VAR1;
IF VAR1 > .Z THEN N + 1;
```

```
IF LAST.VAR100;
FILE PRINT;
PUT VAR100= SUM= N=;
SUM=0;
N=0;
```

E:
```
PROC SQL;
SELECT SUM(VAR1) AS SUM, N(VAR1) AS N FROM TEMP2
GROUP BY VAR100;
```

F:
```
PROC MEANS DATA=TEMP2 SUM N;
VAR VAR1;
BY VAR100;
```

	Operating System			
Test	MVS	Win NT	UNIX	OpenVMS
A: MEANS, with CLASS	100%	100%	100%	100%
B: SQL, unsorted data	140%	165%	153%	49%
C: PROC SORT	119%	349%	209%	153%
D: DATA step	66%	112%	50%	20%
E: SQL, sorted data	106%	108%	114%	27%
F: MEANS, with BY	68%	59%	70%	73%

Notes:

- Obviously, PROC SORT is not an alternative method. It is required for methods D, E, and F. Its CPU time is broken out separately to help you judge the value of sorting the data as well as the best tool to use on sorted data.

- Chapter 5 addresses techniques to speed up PROC SORT, which may influence which strategy is best.

- For method E to out perform method B, the data must not only be sorted. They must be known by the software as being sorted.

For more information, refer to Chapter 5 and the SORTEDBY= data set option.

- Refer to Test 5A for some related results.

Tests Results for Chapter 7: Data Manipulation

Most of these tests run through large DO loops. In general, tests that use DO loops produce more consistent results than tests which read in data sets.

Data manipulation statements execute very quickly. To try to put the savings in context, then, some of the tables below report on entire DATA steps, rather than isolating individual programming statements.

Test 7A: Removing the SUBSTR Function from Assignment Statements

A:
```
DATA _NULL_;
  ORIGINAL='Original String';
  LENGTH FIRST3 $ 3;
  DO I=1 TO 500000;
  END;
```

B:
```
DATA _NULL_;
  ORIGINAL='Original String';
  LENGTH FIRST3 $ 3;
  DO I=1 TO 500000;
    FIRST3=ORIGINAL;
  END;
```

C:
```
DATA _NULL_;
  ORIGINAL='Original String';
  LENGTH FIRST3 $ 3;
  DO I=1 TO 500000;
    FIRST3=SUBSTR(ORIGINAL,1,3);
  END;
```

	Operating System			
Test	MVS	Win NT	UNIX	OpenVMS
A: Baseline DATA step	100%	100%	100%	100%
B: Assignment statement	123%	106%	233%	195%
C: SUBSTR function	651%	296%	648%	445%

Notes:

- Compared to the assignment statement, SUBSTR is exceedingly slow!

- If the test programs were reading in and outputting observations, the DATA steps would be performing more work. In that case, all percentage numbers would be closer to 100% than they are now.

Test 7B: *Removing the SUBSTR Function from Comparisons*

The comparison operator =: tests whether two character strings are equal, based on the number of characters in the shorter string.

```
A: DATA _NULL_;
   ORIGINAL='Original String';
   DO I=1 TO 500000;
   END;
```

```
B: DATA _NULL_;
   ORIGINAL='Original String';
   DO I=1 TO 500000;
      IF ORIGINAL =: 'Software' THEN X=5;
   END;
```

```
C: DATA _NULL_;
   ORIGINAL='Original String';
   DO I=1 TO 500000;
      IF SUBSTR(ORIGINAL,1,8)='Software' THEN X=5;
   END;
```

	Operating System			
Test	MVS	Win NT	UNIX	OpenVMS
A: Baseline DATA step	100%	100%	100%	100%
B: Compare using =:	131%	110%	235%	242%
C: SUBSTR function	659%	316%	651%	521%

Notes:

- Once again, SUBSTR slows down the DATA step considerably.

Test 7C: Getting the Current Date

The raw data file contains 100,000 lines.

A:
```
DATA _NULL_;
   INFILE RAWDATA;
   INPUT;
   DATE=TODAY();
```

B:
```
DATA _NULL_;
   INFILE RAWDATA;
   INPUT;
   IF _N_=1 THEN DATE=TODAY();
   RETAIN DATE;
```

C:
```
DATA _NULL_;
   INFILE RAWDATA;
   INPUT;
   RETAIN DATE "&SYSDATE"D;
```

	Operating System			
Test	MVS	Win NT	UNIX	OpenVMS
A: TODAY(), every obs	100%	100%	100%	100%
B: TODAY(), first obs	64%	32%	68%	86%
C: RETAIN statement	62%	30%	67%	85%

Notes:

- The macro variable &SYSDATE returns the current date in the DATE7. format, such as 01AUG99.

Test 7D: *Eliminating Calls to the LAG Function*

A:
```
DATA _NULL_;
   SET _030VARS (KEEP=VAR1);
```

B:
```
DATA _NULL_;
   SET _030VARS (KEEP=VAR1);
NEWVAR=LAG(VAR1);
```

C:
```
DATA _NULL_;
   SET _030VARS (KEEP=VAR1);
   OUTPUT;
NEWVAR=VAR1;
RETAIN NEWVAR;
```

	Operating System			
Test	MVS	Win NT	UNIX	OpenVMS
A: Baseline DATA step	100%	100%	100%	100%
B: LAG function	110%	108%	108%	92%
C: LAG workaround	102%	112%	101%	100%

Notes:

- Not much impact here. The most surprising number is the 92% under OpenVMS, meaning that adding the LAG function reduced CPU time! Of course, that must be due to random variation in the speed of reading in observations.

- Under MVS and UNIX, the two operating systems that benefited by avoiding calls to the LAG function, the results were consistent. Each of three test runs using method C ran faster than every test run using method B. On the other two operating systems, the CPU time for test runs exhibited more variation.

Test 7E: *Controlling Numeric to Character Conversions*

In the statements below, VALUE is numeric, with a value of 125. C is character, with a length of 3.

A: `C=VALUE;`

B: `C=PUT(VALUE, 3.);`

	Operating System			
Test	MVS	Win NT	UNIX	OpenVMS
A: Software converts	100%	100%	100%	100%
B: PUT function	119%	135%	111%	113%

Notes:

- These percentages compare one statement's CPU time to the other's CPU time. They do not demonstrate the impact on CPU time within the context of a larger DATA step. A complete DATA step would also utilize CPU time to read in observations, perform additional calculations, and output observations. The same applies to all remaining tests except one (7P).

Test 7F: The Order of Nested DO Loops

```
A: DO CUSTOMER=1 TO 100;
      DO PRICE=5 TO 25 BY 5;
         DO QUANTITY=1 TO 3;
            TOTCOST=PRICE*QUANTITY;
         END;
      END;
   END;

B: DO QUANTITY=1 TO 3;
      DO PRICE=5 TO 25 BY 5;
         DO CUSTOMER=1 TO 100;
            TOTCOST=PRICE*QUANTITY;
         END;
      END;
   END;
C: DO QUANTITY=1 TO 3;
      DO PRICE=5 TO 25 BY 5;
         TOTCOST=PRICE*QUANTITY;
         DO CUSTOMER=1 TO 100;
         END;
      END;
   END;
```

	Operating System			
Test	MVS	Win NT	UNIX	OpenVMS
A: Original nesting	100%	100%	100%	100%
B: Revised nesting	75%	65%	60%	63%
C: Shift assignment statement	42%	58%	38%	49%

Notes:

- Program B eliminates 582 DO loops. Program C eliminates 1,485 assignment statements. Except under MVS, eliminating the DO loops saves more.

Test 7G: Test for Division by Zero

In the statements below, NUMER is 1, while DENOM is 0.

A: `QUOTIENT = NUMER/DENOM;`

B: `IF DENOM THEN QUOTIENT = NUMER/DENOM;`

The results of this test were so outrageous that they do not form a meaningful table. The first statement took at least 400 times the CPU time needed for the second (for observations where DENOM is zero). On some operating systems, the ratio was much higher.

The condition IF DENOM is true whenever DENOM is neither zero nor missing.

Test 7H: Test for Missing Values

In these test statements, four variables (A, B, C, and D) are 1, while the last variable (MISSING) is missing.

A: `TOTAL = MISSING + A + B + C + D;`

B: `TOTAL = A + B + C + D + MISSING;`

	Operating System			
Test	MVS	Win NT	UNIX	OpenVMS
A: MISSING first	100%	100%	100%	100%
B: MISSING last	40%	35%	44%	26%

Notes:

- A small change cuts more than half the CPU time for the one statement.

Test 7I: *Multiplication vs. Division*

Both variables (A and B) are equal to 1.

A: `C = A / B;`

B: `C = A * B;`

	Operating System			
Test	MVS	Win NT	UNIX	OpenVMS
A: Division	100%	100%	100%	100%
B: Multiplication	62%	98%	30%	48%

Notes:

- Both statements execute quite quickly. Even when the percentage reduction is large, the absolute savings are small.

- The small savings under Windows NT may constitute a praiseworthy result. Relative to other operating systems, it may be that division runs quickly, not that multiplication runs slowly.

Test 7J: *Grouping Numeric Constants*

A: `C = 1 + 8 + A;`

B: `C = 1 + A + 8;`

	Operating System			
Test	MVS	Win NT	UNIX	OpenVMS
A: Group constants	100%	100%	100%	100%
B: Separate constants	190%	188%	239%	188%

Notes:

- Because the second statement performs twice as much math, it ought to take twice as long to execute.

- These statements run incredibly quickly. To obtain reliable test numbers, each statement ran 2,500,000 times. While the percentage savings may be high, the absolute savings are small.

Test 7K: Adding ELSE to a Series of Comparisons

Embedded comments below indicate which comparison was true for that test.

```
A: IF NAME='WILLIAM' THEN CATEGORY='HUSBAND ';        /* true */
   IF NAME='HILLARY' THEN CATEGORY='WIFE    ';
   IF NAME='CHELSEA' THEN CATEGORY='DAUGHTER';

B: IF      NAME='WILLIAM' THEN CATEGORY='HUSBAND '; /* true */
   ELSE IF NAME='HILLARY' THEN CATEGORY='WIFE    ';
   ELSE IF NAME='CHELSEA' THEN CATEGORY='DAUGHTER';

C: IF      NAME='WILLIAM' THEN CATEGORY='HUSBAND ';
   ELSE IF NAME='HILLARY' THEN CATEGORY='WIFE    '; /* true */
   ELSE IF NAME='CHELSEA' THEN CATEGORY='DAUGHTER';

D: IF      NAME='WILLIAM' THEN CATEGORY='HUSBAND ';
   ELSE IF NAME='HILLARY' THEN CATEGORY='WIFE    ';
   ELSE IF NAME='CHELSEA' THEN CATEGORY='DAUGHTER'; /* true */
```

	Operating System			
Test	MVS	Win NT	UNIX	OpenVMS
A: Three IF THENs	100%	100%	100%	100%
B: IF / ELSE IF, match on 1st try	53%	52%	57%	54%
C: IF / ELSE IF, match on 2nd try	81%	85%	72%	73%
D: IF / ELSE IF, match on 3rd try	108%	98%	97%	97%

Test 7L: IF THEN vs. SELECT

Again, embedded comments below indicate which comparison was true for that test. The baseline tests in method A are the same baselines used above in Test 7K.

```
A:  IF NAME='WILLIAM'  THEN CATEGORY='HUSBAND ';    /* true */
    IF NAME='HILLARY'  THEN CATEGORY='WIFE     ';
    IF NAME='CHELSEA'  THEN CATEGORY='DAUGHTER';

B:  SELECT;
        WHEN (NAME='WILLIAM') CATEGORY='HUSBAND '; /* true */
        WHEN (NAME='HILLARY') CATEGORY='WIFE     ';
        WHEN (NAME='CHELSEA') CATEGORY='DAUGHTER';
        OTHERWISE;
    END;

C:  SELECT;
        WHEN (NAME='WILLIAM') CATEGORY='HUSBAND ';
        WHEN (NAME='HILLARY') CATEGORY='WIFE     '; /* true */
        WHEN (NAME='CHELSEA') CATEGORY='DAUGHTER';
        OTHERWISE;
    END;

D:  SELECT (NAME);
        WHEN ('WILLIAM') CATEGORY='HUSBAND ';       /* true */
        WHEN ('HILLARY') CATEGORY='WIFE     ';
        WHEN ('CHELSEA') CATEGORY='DAUGHTER';
        OTHERWISE;
    END;
```

```
E:  SELECT (NAME);
      WHEN ('WILLIAM')  CATEGORY='HUSBAND ';
      WHEN ('HILLARY')  CATEGORY='WIFE    ';         /* true */
      WHEN ('CHELSEA')  CATEGORY='DAUGHTER';
      OTHERWISE;
    END;
```

	Operating System			
Test	MVS	Win NT	UNIX	OpenVMS
A: Three IF THENs, same as Test 7K	100%	100%	100%	100%
B: SELECT;, match on 1st try	50%	62%	58%	49%
C: SELECT;, match on 2nd try	86%	112%	68%	69%
D: SELECT(NAME);, match on 1st try	202%	182%	89%	54%
E: SELECT(NAME);, match on 1st try	298%	285%	111%	80%

Notes:

- In most cases, SELECT performs as well as a series of IF/THEN/ELSE statements (refer to test 7K as well). The exception, 112% under Windows NT, was not an aberration. Each of the three test runs on both methods (A and C) produced consistent results.

- The biggest shock here is that SELECT(NAME) runs relatively slowly. Intuitively, the opposite should be true because NAME never changes as the software moves through the WHEN statements.

Test 7M: IF/THEN/ELSE vs. the PUT Function

A preliminary PROC FORMAT created the TESTING format as follows:

```
VALUE TESTING  1='A'   2='B'   3='C'   4='D'   5='E'   6='F'   7='G'
               8='H'   9='I'  10='J'  11='K'  12='L'  13='M'  14='N'
              15='O'  16='P'  17='Q'  18='R'  19='S'  20='T'  21='U'
              22='V'  23='W'  24='X'  25='Y'  26='Z';
```

Each test ran inside a DO loop that looped from N=1 to N=26.

```
A:  IF      N= 1 THEN LETTER='A';
    ELSE IF N= 2 THEN LETTER='B';
    ELSE IF N= 3 THEN LETTER='C';
     . . .

    ELSE IF N=25 THEN LETTER='Y';
    ELSE IF N=26 THEN LETTER='Z';

B:  LETTER = PUT(N, TESTING.);
```

| | Operating System | | | |
Test	MVS	Win NT	UNIX	OpenVMS
A: IF/THEN/ELSE	100%	100%	100%	100%
B: PUT function	201%	160%	248%	177%

Notes:

- Again, IF/THEN/ELSE runs incredibly quickly. Presumably, as the number of categories increases, the binary search of the PUT function becomes more and more efficient.

Test 7N: The IN Operator vs. a Logical OR

```
A:  IF SYMBOL IN ('GE', 'GM', 'F') THEN X=1;   /* Symbol is GE */

B:  IF SYMBOL IN ('GE', 'GM', 'F') THEN X=1;   /* Symbol is GM */
```

```
C: IF SYMBOL IN ('GE', 'GM', 'F') THEN X=1;   /* Symbol is F  */

D: IF SYMBOL='GE' OR SYMBOL='GM' OR
      SYMBOL='F' THEN X=1;                     /* Symbol is GE */
E: IF SYMBOL='GE' OR SYMBOL='GM' OR
      SYMBOL='F' THEN X=1;                     /* Symbol is GM */

F: IF SYMBOL='GE' OR SYMBOL='GM' OR
      SYMBOL='F' THEN X=1;                     /* Symbol is F  */
```

	Operating System			
Test	MVS	Win NT	UNIX	OpenVMS
A: IN operator, match on 1st try	100%	100%	100%	100%
B: IN operator, match on 2nd try	171%	170%	137%	161%
C: IN operator, match on 3rd try	259%	240%	174%	202%
D: Series of ORs, match on 1st try	5%	10%	42%	75%
E: Series of ORs, match on 2nd try	9%	13%	64%	149%
F: Series of ORs, match on 3rd try	18%	12%	100%	207%

Notes:

- OpenVMS generates the "quirky" result, where the IN operator runs just as quickly. On all other operating systems, a series of three OR conditions runs much faster.

Test 7O: IF X vs. IF X=1

The value of X is always zero, so both conditions are always false.

A: ```
IF X=5 THEN Y=1;
```

B: ```
IF X   THEN Y=1;
```

	Operating System			
Test	MVS	Win NT	UNIX	OpenVMS
A: Evaluate X=5;	100%	100%	100%	100%
B: Evaluate X	31%	16%	79%	24%

Notes:

- Once again, large percentage savings represent small absolute savings. The test programs ran through 2,500,000 iterations just to accumulate a few seconds of CPU time.

Test 7P: A RETAIN Statement vs. Assignment Statements

Because the absolute differences in CPU time are so small, these test results incorporate the CPU time for a baseline DATA step that reads 100,000 observations.

A: ```
DATA _NULL_;
   SET _030VARS (KEEP=VAR1);
```

B: ```
DATA _NULL_;
   SET _030VARS (KEEP=VAR1);
   A=1; B=1; C=1; D=1; E=1;
```

C: ```
DATA _NULL_;
   SET _030VARS (KEEP=VAR1);
   RETAIN A B C D E 1;
```
```

	Operating System			
Test	MVS	Win NT	UNIX	OpenVMS
A: Baseline DATA step	100%	100%	100%	100%
B: Assign values	106%	119%	110%	104%
C: Retain values	101%	103%	100%	94%

Notes:

- The OpenVMS tests generate inconsistent results. Looking at the numbers, the first test (method A) is probably an overestimate. Logically, method C should always take just slightly longer than method A.

- Surprisingly, the extra cost of method B (vs. method C) is consistently measurable.

Test 7Q: *Array Elements vs. Variables*

A:
```
ARRAY X {5};
X1=0; X2=0; X3=0; X4=0; X5=0;
```

B:
```
ARRAY X {5};
X{1}=0; X{2}=0; X{3}=0; X{4}=0; X{5}=0;
```

	Operating System			
Test	MVS	Win NT	UNIX	OpenVMS
A: Variables	100%	100%	100%	100%
B: Array elements	401%	186%	118%	242%

Efficiency in the Future

As the software evolves, efficiency considerations change. New tools emerge, and existing tools occasionally improve. Consider these two statements, for example:

```
IF X=1 OR X=3 OR X=5 THEN DELETE;
IF X IN (1, 3, 5) THEN DELETE;
```

Although both statements delete the same observations, they execute at different speeds. The IN operator used to run faster. It would search through the list of values until it found a match, but then would stop searching. The first statement, on the other hand, would check every condition. Logically, we know if X=1 that there is no need to check whether X=3 or X=5. But the software would check anyway, making the first statement relatively inefficient. For Release 6.07 in the MVS, VMS, and VMS environments, software improvements allow the first statement to stop checking once it finds a match, therefore running faster than the second statement.

As the software evolves, how will you know which techniques change and which remain most efficient? This book helps in two ways. First, the test programs used in this book are available online (see the inside back cover for information on SAS Online Samples). You can download any or all of them, and run them on your own hardware and your own release of the software. When you install a new release, you will be able to run these programs to see if efficiency considerations have changed. In addition, the first section of this chapter describes how to write your own tests of efficiency. As SAS Institute develops new tools, you can create your own test programs.

Index

O

observations
  repeating single 57-59
  subsetting 20-26, 187-192, 199
observations, selecting test samples
  every Nth observation 35-36
  N observations 32-35
  random samples 36-38
OR (logical) versus IN operator 131, 219-220
output data sets, suppressing 46

P

permanent SAS data sets
  and file handling 53-54
  reading from 29-32
  sort order 73-76, 78
printing, formats for 45-46, 107, 143, 153-154, 193-194
PROC APPEND, appending to files 56
PROC COPY
  copying files 55
  stabilizing SAS data libraries 148-150
PROC DATASETS
  appending to files 56
  copying files 55
  modifying data structures 56-57
PROC FREQ
  and unsorted data 52
  counting data 171-172
PROC FREQ output data sets
  resummarizing 106-107
  structure of 105-106
PROC MEANS
  See also PROC SUMMARY output data sets
  computing statistics 171-172
  versus DATA steps 95-97, 205-208
  versus PROC SUMMARY 97, 204-205

versus SQL 95-97, 205-208
PROC SORT, speeding up 81-83, 202-203
PROC SQL versus
  DATA steps 62-64, 198-201, 205-208
  PROC MEANS 95-97, 205-208
PROC SUMMARY output data sets
  integrating summarization levels 101-104
  resummarizing 104-105
  structure of 98-100
PROC SUMMARY versus PROC MEANS 97, 204-205
PROC TABULATE, summarizing data for 49, 196-197
procedures
  PROC APPEND 56
  PROC COPY, copying files 55
  PROC COPY, stabilizing SAS data libraries 148-150
  PROC DATASETS 55-57
  PROC FREQ, and unsorted data 52
  PROC FREQ, counting data 171-172
  PROC FREQ output data sets 105-107
  PROC MEANS, computing statistics 171-17?
  PROC MEANS versus DATA steps 95-97, 205-209
  PROC MEANS versus PROC SUMMARY 97, 204-205
  PROC MEANS versus SQL 95-97. 205-208
  PROC SQL versus DATA steps 62-64, 198-201
  PROC SQL versus PROC MEANS 95-97, 205-208
  PROC SUMMARY output data sets 98-105
  PROC SUMMARY versus PROC MEANS 97
  PROC TABULATE, summarizing data for 49, 196-197
procedures, replacing with DATA steps 94-95

Call your local SAS® office to order these other books and tapes available through the Books by Users℠ program:

king with the SAS® System

rik W. Tilanus ....Order No. A55190

r Guide to Survey Research Using the
® System

rcher Gravely ....Order No. A55688

Audio Tapes
100 Essential SAS® Software Concepts (set of two)

by **Rick Aster** ....Order No. A55309

A Look at SAS® Files (set of two)

by **Rick Aster** ....Order No. A55207

*Welcome * Bienvenue *Willkommen *Yohkoso * Bienvenido*

SAS® Publications Is Easy to Reach

Visit our SAS Publications Web page located at www.sas.com/pubs/

You will find product and service details, including

- **sample chapters**
- **tables of contents**
- **author biographies**
- **book reviews**

Learn about

- **regional user groups conferences**
- **trade show sites and dates**
- **authoring opportunities**
- **custom textbooks**
- **FREE Desk copies**

Order books with ease at our secured Web page!

Explore all the services that Publications has to offer!

Your Listserv Subscription Brings the News to You Automatically

Do you want to be among the first to learn about the latest books and services available from SAS Publications? Subscribe to our listserv **newdocnews-l** and automatically receive the following once each month: a description of the new titles, the applicable environments or operating systems, and the applicable SAS release(s). To subscribe:

1. Send an e-mail message to **listserv@vm.sas.com**

2. Leave the "Subject" line blank

3. Use the following text for your message:

   subscribe newdocnews-l *your-first-name your-last-name*

   For example: subscribe newdocnews-l John Doe

   Please note: newdocnews-l ◄——— that's the letter "l" not the number "1".

For customers outside the U.S., contact your local SAS office for listserv information.

Create Customized Textbooks Quickly, Easily, and Affordably

SelecText™ offers instructors at U.S. colleges and universities a way to create custom textbooks for courses that teach students how to use SAS software.

For more information, see our Web page at **www.sas.com/selectext/**, or contact our SelecText coordinators by sending e-mail to **selectext@sas.com**.

You're Invited to Publish with SAS Institute's User Publishing Program

If you enjoy writing about SAS software and how to use it, the User Publishing Program at SAS Institute Inc. offers a variety of publishing options. We are actively recruiting authors to publish books, articles, and sample code. Do you find the idea of writing a book or an article by yourself a little intimidating? Consider writing with a co-author. Keep in mind that you will receive complete editorial and publishing support, access to our users, technical advice and assistance, and competitive royalties. Please contact us for an author packet. E-mail us at **sasbbu@sas.com** or call 919-677-8000, then press 1-6479. See the SAS Publications Web page at **www.sas.com/pubs/** for complete information.

Read All about It in *Authorline*®!

Our User Publishing newsletter, *Authorline*, features author interviews, conference news, and informational updates and highlights from our User Publishing Program. Published quarterly, *Authorline* is available free of charge. To subscribe, send e-mail to **sasbbu@sas.com** or call 919-677-8000, then press 1-6479.

See *Observations*®, Our Online Technical Journal

Feature articles from *Observations*®: *The Technical Journal for SAS*® *Software Users* are now available online at **www.sas.com/obs/**. Take a look at what your fellow SAS software users and SAS Institute experts have to tell you. You may decide that you, too, have information to share. If you are interested in writing for *Observations*, send e-mail to **sasbbu@sas.com** or call 919-677-8000, then press 1-6479.

Book Discount Offered at SAS Public Training Courses!

When you attend one of our SAS Public Training Courses at any of our regional Training Centers in the U.S., you will receive a 15% discount on any book orders placed during the course. Each course has a list of recommended books to choose from, and the books are displayed for you to see. Take advantage of this offer at the next course you attend!

SAS Institute Inc.
SAS Campus Drive
Cary, NC 27513-2414
Fax 919-677-4444

E-mail: sasbook@sas.com
Web page: www.sas.com/pubs/
To order books, call Book Sales at 800-727-3228*
For other SAS Institute business, call 919-677-8000*

* Note: Customers outside the U.S. should contact their local SAS office.